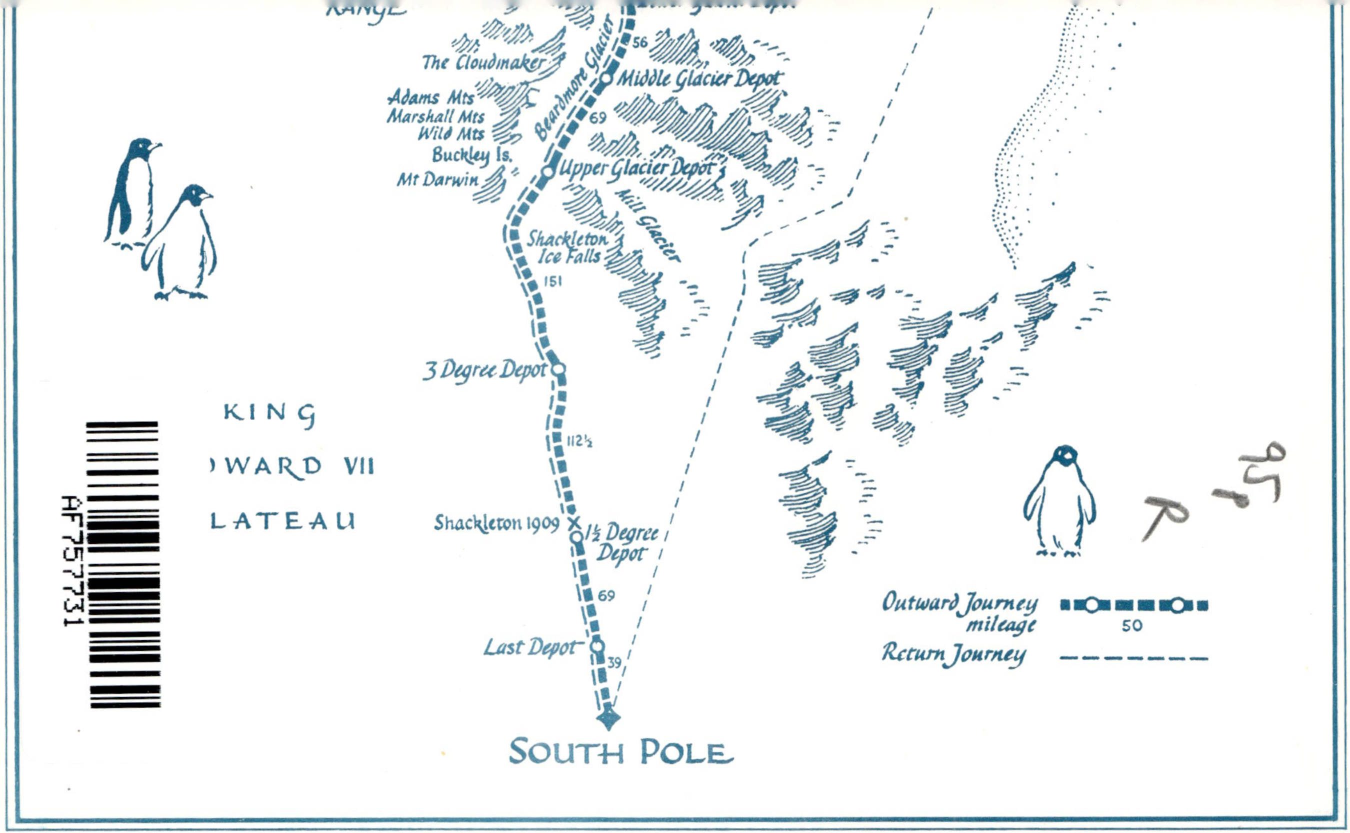
The Cloudmaker
Beardmore Glacier
56
Middle Glacier Depot
Adams Mts
Marshall Mts
Wild Mts
Buckley Is.
Mt Darwin
69
Upper Glacier Depot
Mill Glacier
Shackleton Ice Falls
151
3 Degree Depot
KING
WARD VII
LATEAU
112½
Shackleton 1909
1½ Degree Depot
69
Last Depot
39
SOUTH POLE
Outward Journey
mileage
50
Return Journey

THE NEW WINDMILL SERIES
General Editors: Anne and Ian Serraillier

59

THE EPIC OF CAPTAIN SCOTT

A straightforward unvarnished account of Captain Scott's tragic and inspiring last journey, written by an author familiar at first hand with Arctic conditions. The illustrations, many of them by Herbert Ponting, the expedition's photographer, are superb.

The Author

Sir Martin Lindsay was a member of the British Arctic Air-Route Expedition, 1930–31.

In 1934 he led the British Trans-Greenland Expedition which made the longest self-supporting sledge journey in history, of 1050 miles (*Guinness Book of Records*), to be the first to reach and map some 350 miles of the highest mountains within the Arctic Circle. The two principal ranges he named Crown Prince Frederik of Denmark Mountains and Prince of Wales Mountains.

MARTIN LINDSAY

The Epic of Captain Scott

HEINEMANN EDUCATIONAL BOOKS
LONDON

Heinemann Educational Books Ltd
22 Bedford Square, London WC1B 3HH
LONDON EDINBURGH MELBOURNE AUCKLAND
HONG KONG SINGAPORE KUALA LUMPUR NEW DELHI
IBADAN NAIROBI JOHANNESBURG KINGSTON
EXETER (NH) PORT OF SPAIN

ISBN 0 435 12059 x

First published 1933
First published in the New Windmill Series 1962
Reprinted 1965, 1966, 1968, 1970, 1973, 1977
Reprinted with new Epilogue 1980

Printed and bound in Great Britain by
Morrison & Gibb Ltd, London and Edinburgh

Acknowledgment

My grateful thanks are due to Lady Hilton Young, for allowing me to quote certain passages from Captain Scott's diary and letters. A debt of gratitude is also owing to Mr Apsley Cherry-Garrard for permitting me to quote from *The Worst Journey in the World*, from which most of the other diary passages are taken, and to Professor Debenham of the Scott Polar Research Institute and a member of Captain Scott's last expedition, for much helpful advice.

To Mrs E. A. Wilson, Lady Maxwell, Sir James Barrie, Mr Herbert G. Ponting, and Messrs John Murray, all of whom placed copyright material at my disposal, I also say thank you.

1 October 1933 — Martin Lindsay

Note to New Windmill edition

Thanks are due to the following for permission to reproduce the photographs in this edition:

The Scott Polar Research Institute and Mr L. M. Forbes
Trans-Antarctic Expedition
Miss Edith M. J. Bowers
British Museum
Paul Popper Picture Agency

Note on 1980 reprint

For this reprint the author has rewritten the Epilogue.

Contents

Illustrations

Between pages 64 and 65

Maps

Prologue

MANY YEARS have passed since Captain Robert Falcon Scott and the whole of his party perished from starvation and exhaustion on their way back from the South Pole. Petty Officer Edgar Evans was the first to die, and the others struggled onward for three hundred miles. Then, crippled by scurvy and frostbite, Captain L. E. G. Oates walked out to his death in a blizzard in order not to delay his companions. Captain Scott, Dr Wilson, and Lieutenant Bowers fought on for another four days, and then camped with but two meals left. If the weather had held they could have got through the last thirteen miles to One Ton Depot, where plenty awaited them. But again a blizzard descended upon them. It lasted for nine days. It was the end.

There are few events in history to be compared with the final tragedy enacted in that silent desert of snow – the dying leader, with the dead bodies of his dearest friends beside him, writing, writing until the pen slipped from his fingers, with never a thought for himself, seeking only to give consolation to others.

'*We took risks, we knew we took them; things have come*

out against us, and therefore we have no cause for complaint, but bow to the will of Providence, determined still to do our best to the last. But if we have been willing to give our lives to this enterprise, which is for the honour of our country, I appeal to our countrymen to see that those who depend on us are properly cared for.

'Had we lived, I should have had a tale to tell of the hardihood, endurance, and courage of my companions which would have stirred the heart of every Englishman. These rough notes and our dead bodies must tell the tale . . .

They conquered, and then met defeat. But never was death more swallowed up in victory. They died, but they are not forgotten. We who remember give homage.

I

Introduction

The Romantic Voyages

THE ANCIENT Greeks discovered that the world is round. They made the first globe, and on it sketched a southern hemisphere to balance that of the north. The north they called *Arktos*, giving it the name of the Bear whose outline they could trace in the heavens above it. *Antarktos* was therefore the only possible name for the land that they assumed lay opposite – hence *Antarctic*.

To complete the globe the early geographers relied largely upon analogy and tradition. The Phoenician sailormen were of little assistance. They reported that the southern temperate zone, to say nothing of Antarktos beyond, could not be reached because of the unendurable heat of the Equator, where no life was possible; that the seas were alive with whales of incredible size and monsters defying description; that the waves rose to mountain height; and that the Devil himself stretched a hand out of the sea to snatch up wandering ships. Medieval maps, indeed, showed the island *de la man de Satanaxio* in the Atlantic. The early globes varied vastly in detail, for, where legend and the Phoenicians failed, the cartographer had to draw upon his imagination.

In A.D. 43 that dreamer Pomponius Mela gave the name of Antipodeans to the inhabitants of the imaginary land to the south, of whom many fantastic accounts were subsequently written. On his map he drew its outline in continuation of Africa, and thereafter every southerly landfall was mistaken for the unknown continent of Antarktos. Tierra del Fuego, the Solomon Islands, New Guinea, Australia, and New Zealand were all part of this land until the next mariner disproved it.

The existence of Antarctica was never really doubted, although successive voyages failed to locate it. Finally, in 1772, the British Admiralty sent out Captain Cook with orders to settle the problem. The first man to cross the Antarctic Circle, he forced his ships far down into the pack-ice, and found nothing ahead but snow and ice. He proved that the Antipodeans did not exist, for if there were any great continent to the south it could have no connection with lands already known and must be uninhabited.

In the course of his voyages Captain Cook discovered the Isle of Georgia (a name which by the carelessness

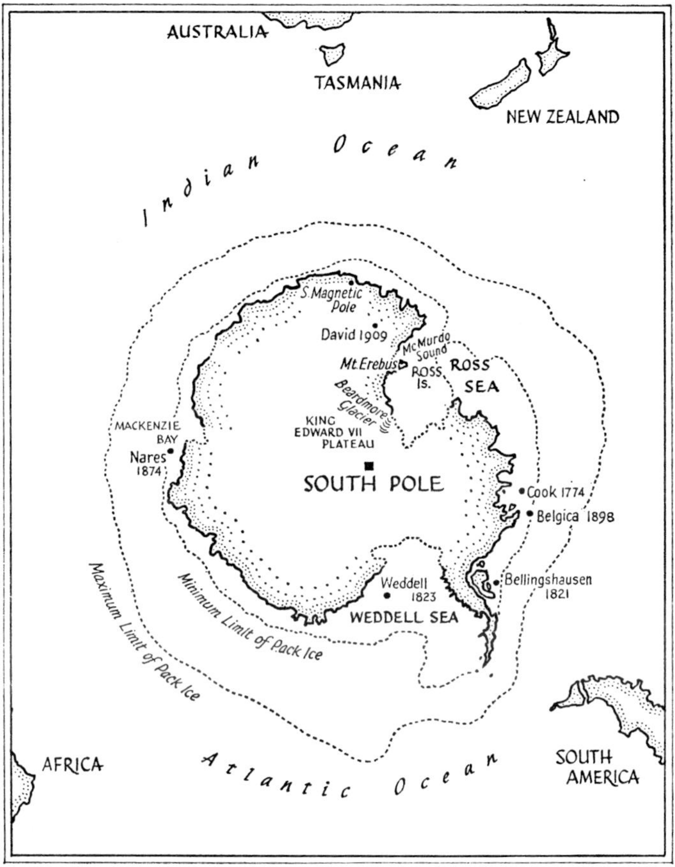

AUSTRALIA
TASMANIA
NEW ZEALAND
Indian Ocean
S. Magnetic Pole
David 1909
McMurdo Sound
Mt. Erebus
ROSS Is.
ROSS SEA
Beardmore Glacier
KING EDWARD VII PLATEAU
MACKENZIE BAY
Nares 1874
SOUTH POLE
Cook 1774
Belgica 1898
Bellingshausen 1821
Weddell 1823
WEDDELL SEA
Minimum Limit of Pack Ice
Maximum Limit of Pack Ice
AFRICA
Atlantic Ocean
SOUTH AMERICA

Antarctica

of an engraver was changed to South Georgia), and gave it as his opinion that the island could never be of any value. The wisest man of his time thus proved his fallibility. In a few years scores of sealers were using it as their base, and it was the pursuit of seal-hunting that made the next inroad into the unknown. Many naval officers were thrown out of employment at the close of the Napoleonic wars, and the most adventurous of them turned to Antarctic sealing; they were fine navigators, able to chart their discoveries and to give a good account of them. Men like Weddell, Biscoe, and Balleny reached high southern latitudes and found new land. A Russian named Bellingshausen made a great pioneering voyage, but the reports of his work did not appear in any other language than his own for many years.

Then came the cause of science. The increase of shipping between Australia, China, and South America necessitated a better knowledge of the magnetic conditions of the South Atlantic and Pacific, and terrestrial magnetism began to be more closely studied. The Royal Society induced the Government to send out two ships, the *Erebus* and *Terror*, to discover the South Magnetic Pole. The leader was James Ross, who had already visited the North Magnetic Pole. Early in 1841 Ross pushed his way through the pack-ice into the sea which now bears his name. Here he discovered a mountainous coast-line, extending to Lat. 78° S., which he called Victoria Land. The Magnetic Pole was located far in its interior. The coast turned eastwards at an inlet which he named McMurdo Sound. Overlooking the bight were two great volcanic peaks, one active and one extinct, to which, with singular appropriateness, the ships stood godmother, and Mount Erebus and Mount Terror were left to mark the most southerly point reached. From here a huge wall of ice, apparently afloat, stretched to the east for 400 miles, and as it barred his way to the Pole Ross christened it the 'Barrier'. The ship's company was enormously impressed by the Barrier, for it was quite unlike anything they had ever heard of, and its origin was a mystery.

It is impossible, in this brief outline, to give any idea of the difficulties and hardships of the early explorers who made these romantic voyages. One cannot read their simple narratives without being struck by the courage and determination displayed; time after time they sailed boldly in among the pack-ice in their small, ill-found, leaking ships, while scurvy, overwork, and under-nourishment played havoc among the crews. Yet, somehow, they struggled on in the face of every conceivable obstacle and did not turn from their course unless obliged to by overpowering necessity.

Then, for nearly thirty years, until the celebrated voyage of the *Challenger*, no ship broke the silence of the South. At last, in 1872, the Admiralty despatched Captain Nares, R.N., with a scientific staff of six men to 'study the depths and the living creatures of all the oceans in the world'. Although Antarctic exploration was not part of the scope and intention of the expedition, Nares made a dash across the Circle and introduced steam-power to those waters. No land was seen, but dredgings were made, and amongst the specimens secured were a number of continental rocks which the ice had undoubtedly carried there from some unknown, unseen land beyond them. The importance of this expedition to Antarctic exploration lay less in the discoveries it made than in the interest it aroused among men of science. But it was another twenty years before much was done.

At the Sixth International Geographical Congress, in London in 1895, a resolution was adopted: 'That the Congress records its opinion that the exploration of the Antarctic Regions is the greatest piece of geographical exploration still to be undertaken. That, in view of the additions to knowledge in almost every branch of science which would result from such a scientific exploration, the Congress recommends that the scientific societies throughout the world should urge, in whatever way seems to them most effective, that this work should be undertaken before the close of the century.' This may be regarded as the inauguration of a great campaign, for during the thirty-eight years between then and the writing of this book there were no fewer than thirteen major expeditions to the Antarctic.

The first of these was the personal effort of an

officer in the Belgian navy, Lieutenant de Gerlache, who in 1897 had the distinction of commanding the first *scientific* Antarctic expedition. After much valuable work had been done on the west coast of Graham Land, the *Belgica* proceeded south-west as far as Lat. 70° 20′ S. and Long. 85° W. At that point a gale opened up the pack, and de Gerlache, in spite of the protests of his staff, let his ship drive southwards into the ice at a time of year when all previous ships had been hurrying to the north. It was a brave but rash decision, and might well have ended in disaster. Lat. 71° 30′ S. was reached, and then the ship was beset for thirteen months. The usual scientific observations were continued and were of considerable importance as none had ever been taken so far south in winter. The Antarctic night proved a terrible ordeal. There was no lack of food, but most of it was in the form of specially prepared extracts, patent nourishments that did not nourish, and the whole company fell ill. One of the scientists died. Damp and extreme cold, together with the ceaseless crashing of the ice all through the endless night, preyed on the sick men's nerves to such an extent that several became temporarily deranged. 'We lived in a madhouse,' wrote one of them. Eventually the *Belgica* escaped from the pack-ice and returned to civilisation with a collection of data which was, for a time, unique.

Meanwhile, through the generosity of Sir George Newnes, Mr Borchgrevink equipped the *Southern*

Cross and sailed for the Ross Sea. A hut was erected at Cape Adare, and the party wintered in very different circumstances from those of the *Belgica* the preceding year. Although the station was subject to violent hurricanes, the general health was satisfactory. The ship returned the following summer, and a southerly course was set along the coast of South Victoria Land. With steam-power the Expedition was able to approach the Barrier much more closely than had been possible in the case of the *Erebus* and *Terror*, and found that it had receded about thirty miles during the sixty years that had elapsed since Ross had discovered it. A landing was made in Long. 164° W., and Mr Borchgrevink and Lieutenant Colbeck made a short sledge trip as far as Lat. 78° 50′ S., thus establishing a 'record'. This journey was important in that it revealed the nature of the Barrier and pointed a way to the South Pole.

The nineteenth century had proved the existence of a great southern continent, but had done little to unveil it. It had, however, made clear how woefully incomplete was our knowledge of the Antarctic. Addressing the Royal Geographical Society in 1893, the Duke of Argyll remarked that more was known about the planet Mars than about a large area of our own globe. However, the efforts of the *Challenger*, the *Belgica*, and the *Southern Cross* had drawn the attention of scientific minds all over the world to the problems that were awaiting solution in the South.

There was scarcely a branch of science to which Antarctica did not offer vast possibilities: to geologists it promised in large measure to complete the story of the earth; to physiographers it could be more useful than any other continent, for there they would see in daily and hourly operation the conditions which had existed over the whole world in past ice-ages; to biologists it might be of paramount importance in view of the significance which attaches to the life of the sea in the evolutionary problem; to meteorologists it was becoming plain that this polar continent in all probability influenced the weather conditions of the whole Southern Hemisphere; to magnetism the importance of the unexplored continent was only equalled by the mystery in which the whole subject was shrouded.

And it was to study all these different scientific problems that, in the first year of the new century, the great British Antarctic Expedition set out for the South.

The 'Discovery'

The British Antarctic Expedition, better known as the 'Discovery Expedition' from the name of the ship which carried it, owed its existence largely to the persistence of one man. Sir Clements Markham had decided many years earlier that a great British expedition should be sent to the Antarctic, but it was not until 1893, when he was elected to the Presidency of the Royal Geographical Society, that he was able to put his plans into effect.

Thirteen years before the expedition took place Sir Clements had selected the leader. In 1887 he was staying on board ship with his cousin Sir Albert Markham, the Arctic explorer, then Commodore of a squadron of the fleet. Already on the look-out for the leader of an Antarctic expedition, he made himself the personal friend of all the midshipmen in the four ships. And there, amongst them, was Robert Falcon Scott.

It was a cutter race which first drew Sir Clements's attention to Scott. The race tested many of the qualities of the competitors, and for some time it was a close thing between two midshipmen, Scott and Hyde Parker. Scott, aged 18, won the race, and a few days later dined with Sir Clements, who was particularly struck with his intelligence and the marked charm of his manner. Sir Clements knew that it would be some years before an expedition could take place, but he believed that Scott was destined to command it. A few years later the two were thrown together again when Scott was torpedo lieutenant of the *Empress of India*, and Sir Clements was 'more than ever impressed by his evident vocation for such a command'.[1]

One day in June 1899, Scott, spending a short leave

1. Markham, *The Lands of Silence*, p. 447.

in London, met Sir Clements Markham and accompanied him home. Little did he then think that twelve years earlier, when he was a midshipman, that far-sighted man had chosen him to lead a polar expedition. That afternoon Scott heard for the first time that there was a prospective Antarctic expedition. Two days afterwards he wrote applying to command it.

A year later, in June 1900, the appointment was made public, and Scott was promoted to the rank of Commander. His education had not specially prepared him for what was to be his life's work. He had had no experience of travelling over snow and ice, nor had he ever tried climbing or ski-ing. He was not a man of science who had studied the problems of the polar regions, nor even the literature of Arctic and Antarctic exploration. He was, however, a remarkably capable young officer who appeared to have a great future in the Navy.

The Expedition was due to sail in a year. Very little had been done except that the building of the ship had been begun in March 1900, for which Scott had not been in any way responsible. He now had to take charge of and hasten all the preparations, at the same time studying from the beginning the history of polar exploration, the technique of polar travelling, and all the different practical and scientific problems. Few men are capable of leading a major polar expedition; fewer still could have done what Scott did in those twelve months before the *Discovery* sailed.

The *Discovery* – one of the last of the stout wooden ships that were once the main defence of Britain – left New Zealand on Christmas Eve, 1901, and reached the Ross Sea on 7 January. A landing was made at Cape Adare, and at several places on the coast of Victoria Land. The ship steamed for 500 miles along the edge of the Barrier, and further than Ross had been able to go in 1842. Beyond the extreme end of the Barrier they discovered King Edward VII Land, but were unable to reach the shore. The pack was thick and the season well advanced, so the *Discovery* returned to McMurdo Sound and anchored opposite a tongue of land now known as Hut Point. The ship was frozen in and used as a base throughout the winter.

The first autumn was spent in several short journeys of discovery which were invaluable for the experience that was gained. Conditions and equipment were new, and, since everybody was equally ignorant, the first sledging parties had to learn by the process of trial and error. We read of men not knowing how to put up their tents, or assemble the cooker, or even how to put on their clothes. A sailor named Vince lost his life through trying to get from his tent to the ship in a blizzard. The Antarctic explorers of 1902 were pioneers and had to pay for their experience; the price might well have been higher. During the winter the expedition set to work to reorganise their equipment and remedy their mistakes. The failures of the autumn had shown that 'food, clothing, everything was wrong, the whole system was bad'.[1] That they rose to the occasion is proved by the successes of the two following summers.

On 22 November 1902, Scott, accompanied by Dr Wilson and E. H. Shackleton, set out on his first Southern venture. The annals of polar exploration contain no greater names than those of these three men, all of whom subsequently perished in quest of geographical discovery. Their journey was not only heroic but historic, for it was the first of a series of similar endeavours.

Dogs were taken to draw the sledges, but they were indifferent specimens and their food was either tainted or of the wrong kind; one after another they all died or had to be killed. The men themselves pulled in harness and were frequently obliged to relay – taking on half the load and then returning for the remainder. On 21 December Wilson told Scott that Shackleton had symptoms of scurvy, and on 30 December, in Lat. 82° 17′ S., the party turned back. By the middle of January Shackleton was seriously ill; the scurvy was much advanced and he was vomiting blood. All three men were weakened by their inadequate rations, but, by putting their sick comrade on the sledge when he was unable to walk, Scott and Wilson got him back alive. On 3 February the party reached the ship after having travelled 400 miles into the unknown South, and having covered altogether 960 miles in 94 days.

1. Scott, *The Voyage of the 'Discovery'*, vol. i, p. 273.

Scott and Wilson were both ill with scurvy and utterly exhausted. It was a magnificent journey which achieved, under arduous conditions, the greatest advance up to that time towards either the North Pole or the South Pole. Three hundred and fifty miles of mountainous coast-line, forming the western boundary of the Great Ice Barrier, had been discovered.

The second winter passed very much like the first, and as soon as spring came sledging was resumed. This season Scott, with two seamen, Edgar Evans and Lashly, made an even better journey to the west. They climbed the Ferrar Glacier to the ice-cap, where they experienced temperatures of *minus* 40° F. although it was the beginning of the summer. A supporting party left them, and the three men continued westwards over the flat plateau which we now know extends beyond the Pole, and probably to the other side of the continent. They pulled their sledge over the dreary wastes of snow for 200 miles, with no landmark to break the monotony, and then turned homewards. The return journey was hazardous as the navigation tables had been blown away in a blizzard, and Scott had to rely upon memory. However, they got back safely, having averaged 15.4 miles a day for a distance of 1098 miles, with a climb of nearly 10,000 feet. Scott justly claimed that this journey came near the limit of possible performance.

Altogether about three thousand miles were

sledged, and excellent scientific work was done in every branch. The most important discovery was that of the great ice plateau, which in itself would have made Scott famous. Never before had a polar expedition returned with such a rich harvest of results.

Shackleton

Ernest Shackleton left the Antarctic, after his initiation with Scott, convinced that one more effort would carry very much further the work that had been begun so well. In 1907 he sailed from England once more, this time in command of his own expedition, and established his base in McMurdo Sound. The aim of the Expedition was to reach both the Geographical and Magnetic Poles.

In the lesser of these two ambitions Shackleton was completely successful. Three of the party, David, with Mawson and Mackay, sledged up the Reeves Glacier and over the plateau to the Magnetic Pole, which they fixed in Lat. 72° 25′ S., Long. 155° 16′ E. A large part of the time on only half rations, they hauled their sledges 1260 miles in 122 days and covered 300 miles of new ground. They brought back with them a mass of glaciological and geological information. It was one of the greatest journeys in the history of exploration.

Meanwhile Shackleton failed by a narrow margin to reach the South Pole in the course of a journey that cannot fairly be described as anything less than brilliant. He had started at the beginning of November 1908 with Adams, Marshall, and Wild, and four ponies. They steered due south instead of bearing towards the western mountains as Scott had done, and kept this course until stopped by the mighty coastal range which stretched across their path. Searching for a way through to the South they discovered the greatest valley glacier in the world and fought their way up it. The Beardmore Glacier is more than twenty miles wide and more than twice the size of any known glacier, and up this mighty highway they struggled to the ice-cap which was 7000 feet above and 100 miles distant from the foot of the glacier. Like the plateau of South Victoria Land the summit proved to be a barren waste of snow.

The party pressed on southwards for a fortnight after passing out of sight of land, and then, when they had reached the high Latitude of 88° 23′ S. and were only 111 miles from the Pole, they were forced to turn back through lack of provisions. A little more food would have enabled them to reach their goal, instead of having to turn for home with the bare minimum with which to get back. The return journey was a race with death.

On the outward journey, when only 38 miles from Hut Point, the polar party had had to reduce their already inadequate ration. They were soon very hungry, and with the prospect of remaining so, as Shackleton wrote, 'for another three months'. The ascent of Beardmore Glacier was extremely hard work, and when they reached the plateau on 19 December, they were suffering acutely from hunger. By 1 January they were beginning to feel weak from starvation. Their ration at this time was scarcely twenty ounces a day, about half what they needed. Added to this hardship, when on the plateau between 8000 and 10,000 feet above sea-level, they encountered winds of a velocity from 80 to 90 miles an hour, and temperatures that fell to *minus* 40° F.

When they got back to their base they were ill with dysentery and had endured *fifteen and a half weeks'* semi-starvation. They had done great deeds on less food than any other sledging party that has not perished. But they had not reached the Pole.

The 'Terra Nova'

As soon as the Discovery Expedition had been wound up, Captain Scott resumed the duties of a naval officer. He spent the next four years in command of battleships and in confidential work at the Admiralty. But while he paced the bridge of his ship or sat at a desk in Whitehall, his thoughts often wandered to the far-distant South where he was longing to return to complete his work.

In 1910 he launched his last venture, the main purpose of which was to reach the South Pole. This was understood from the beginning by the public and the subscribers of funds. It is said that Scott was actually more interested in the scientific programme, and certainly he arranged for another season's work after the Pole would have been reached. Dr Wilson, Scott's former sledging companion, was in charge of the scientific staff, which was on a scale without precedent: instead of the two biologists of the *Discovery* there were four; instead of one geologist there were three, one of them being a specialist in physiography; and instead of one physicist there were two. Of the twelve seamen to be landed, four of them were old *Discovery* hands, and three of these were to be closely concerned with the Mighty Epic.

This time Scott did not take over a going concern, but had to organise each department and beg for every penny of the funds. In September he bought the *Terra Nova*, 744 tons, the largest of the Dundee whalers, and had her refitted in the West India Docks. The Expedition left England in June 1910, and reached Melbourne on 12 October. There a dramatic telegram was waiting for Scott:

'Madeira. Am going south. – AMUNDSEN.'

This message was, to say the least of it, disturbing,

for it meant that they were in competition with a very big man. Captain Roald Amundsen was one of the most famous explorers of the time, and in 1905 had achieved the ambition of centuries by passing through the North-West Passage. Having been in the Antarctic before Scott, with the *Belgica* Expedition of 1897–99, Amundsen was under no obligation to consider the South Pole in any sense British property.

The *Terra Nova* left civilisation on 29 November, and two days later almost sank in a gale. She entered the pack-ice on 9 December and, unlike the more fortunate *Discovery*, took three weeks to get through. Cape Crozier was sighted on 3 January. Wilson was very anxious for their base to be established there as it would have been nearer the Pole, but no suitable beach could be found. A site was therefore chosen on Cape Evans, 18 miles north of Scott's former quarters at Hut Point, and landing was started immediately. In a week the timber, stores, fuel, ponies, and equipment were all on shore. In a fortnight the hut was built. In three weeks they were ready to begin sledging.

An important journey was now embarked upon in order to leave about a ton of sledging stores and pony fodder in a depot on the Barrier. This cache was to be placed on the route to the Pole and as far south as was possible before the winter set in. Scott himself led a party of twelve men. From Safety Camp – 2½ miles from the end of the Barrier, and so called because it was considered to be beyond the reach of any dislocation of ice at the edge – they marched for three days E.S.E. to avoid the crevasses that radiate from a nunatak called White Island. Then, at a point which they christened Corner Camp, they turned due south and travelled on to Lat. 79° 29′ S., where a cairn was built and 2181 lb. of provisions were deposited. This was called One Ton Depot, and was 131 miles from Hut Point. The return journey was not without adventure, both among crevasses and on the sea-ice which broke up when part of the cavalcade was crossing it.

Meanwhile the *Terra Nova* had sailed eastwards in order to try to leave an independent party somewhere on King Edward VII Land. In the course of the voyage they visited the Bay of Whales, and there, to their great surprise, they found Nansen's old ship, the *Fram*. Amundsen, with a hard-looking lot of men, mostly expert ski-runners, was at work building his base in a position 69 miles nearer the Pole than McMurdo Sound. Before returning to New Zealand the *Terra Nova* called in at Cape Evans and left this unwelcome piece of news to greet Scott on his return from the depot-laying journey.

Scientific work was carried on at Cape Evans throughout the winter. A cavern was dug in a block of ice for the differential magnetic equipment; a hut was built for absolute magnetic observations, and the meteorological instruments were placed on a small hill behind it. A truly fantastic journey in the darkness of mid-winter was undertaken by Wilson, Cherry-

Garrard, and Bowers to Cape Crozier in order to obtain emperor-penguin embryos. The temperature fell as low as *minus* 79° F., their tent was blown away in a gale, and they could hardly have survived if Bowers had not succeeded in finding it among some boulders half a mile away. It was a miracle. Finally, having endured incredible hardships on the worst journey in the world, they got back to Cape Evans on 1 August.

From that time forward all was preparation for the great Southern Journey. The distance from Cape Evans to the Pole was 922 miles, and Scott had three transport units – motor sledges, ponies, and dog-teams – upon which to base his organisation. The plan was for the two motor sledges, which were still in the experimental stage and not to be relied upon, to travel as far across the Barrier as they could, and for the nine ponies and two dog-teams to go to the foot of the Beardmore Glacier, where the former would be killed for fresh food and the latter would return home. From the foot of the glacier three man-hauling teams were to go forward: two of them would act as supporting parties and turn back, the first near the head of the Beardmore and the second at a point on the plateau from which the final party should be able to go on unaided to the Pole. Eight depots were to be placed along the route, and the most elaborate calculations were made about the quantity of food to be left at each depot and the quantity to be taken by each returning party. A start could not be made before about 1 November, for any earlier would be too cold for the ponies. Scott had long realised that they could not race the Norwegians' dog-teams.

October, with its final preparations, its last-minute anxieties and unbearable suspense, dragged slowly out the hundred days. And on the last of them Scott wrote in his diary:

'The future is in the lap of the gods; I can think of nothing left undone to deserve success.'

II

The Mighty Epic

SCOTT WROTE: 'The Southern Journey involves the most important object of the Expedition. . . . One cannot afford to be blind to the situation: the scientific public, as well as the more general public, will gauge the result of the scientific work of the Expedition largely in accordance with the success or failure of the main object [the Polar Journey]. With success all roads will be made easy, all work will receive its proper consideration. With failure even the most brilliant work may be neglected and forgotten, at least for a time.'

Organisation

Motor Party

Lieut. E. R. G. R. Evans, R.N.
Mr B. C. Day.
W. Lashly, Chief Stoker, R. N.
F. J. Hooper, Steward, late R.N.

Pony Party

Capt. R. F. Scott, C.V.O., R.N.
Dr E. A. Wilson.
Capt. L. E. G. Oates, Inniskilling Dragoons.
Lieut. H. R. Bowers, Royal Indian Marines.
Surgeon-Lieut. E. L. Atkinson, R.N.
Mr A. Cherry-Garrard.
Mr C. S. Wright.
Edgar Evans, Petty Officer, R.N.
Patrick Keohane, Petty Officer, R.N.
Thomas Crean, Petty Officer, R.N.

Dog-Teams

Mr C. H. Meares.
Dimitri.

ONE

The Barrier

(421 miles: 27 Oct.—9 Dec.)

THE TRYING days of waiting were over. At last the time had come. Their plans had been brought to maturity, and the preparations of years and months were finished for better or for worse. It was up to *them* now.

To allow them time to make repairs, if necessary, the motor party started first. Their orders were to go only as far as Lat. 80° 30′ S. and there to camp and wait for the others. The first fifty miles of the polar journey were bad going for the ponies as the sea-ice was hummocky and deep snow was usual near the edge of the Barrier. So the motor sledges might be very useful, enabling the ponies to march light until the surface improved.

In charge of the party was Lieutenant 'Teddy' Evans (not to be confused with Petty Officer Edgar Evans), later famous as 'Evans of the *Broke*'. With him were Day, who had been Shackleton's engineer in 1908, and two seamen, Lashly and Hooper. Lashly had already proved his worth on Scott's plateau journey of 1903. They left Cape Evans on 24 October 1911, Evans and Day with the first motor and Lashly and Hooper with the second. Everybody turned out to cheer them on their way, and many were the jeers at the despised motorists. But this derisive attitude hid the most sincere good wishes, for every mile the motors went was so much to the good.

Progress over the sea-ice was poor as the tracks could not get a proper grip on the slippery surface. In three and a half hours they had only covered three and a half miles; it seemed impossible for them to go any faster. In the afternoon conditions became even worse; after being pulled and pushed for hours the motors were only one mile farther on, and the party camped at 10 p.m., thoroughly worn out. For the first day of an eighteen-hundred-mile journey it was scarcely encouraging.

It took them three days to reach Hut Point, only eighteen miles from Cape Evans. The old *Discovery* quarters always proved very useful on the way to the Barrier, and here Evans's party spent their last night in comfort for many a week.

In the morning the whole party went down to send off the motor sledges. After warming up the carburettor with a blow-lamp, and making a few minor adjustments, the engines started, and in bursts of half a mile they soon ran down the short distance to the edge of the Barrier. Now came the test that everybody had secretly been dreading. Would the sledges climb the slope?

Day got into position opposite a convenient snow ramp; Scott looked anxious; the men who were to push prepared to do their uttermost. Then with a whirr the first motor started. The wooden soles of the tractor-belt clattered on the ice as the sledge ran across the tide-crack. It paused a fraction of a second, and then at full speed ran on and up the slope. Almost before they knew it the first motor stood on the Great Ice Barrier. Everybody cheered. Petty Officer Evans said: 'Lord, sir, I reckon if them things can go on like that you wouldn't want nothing else.'

Then Lashly followed and, although he was not quite so dashing on account of the lower speed of his machine, he too climbed on to the Barrier without difficulty. One last hurried handshake, and Scott's party turned back for Hut Point. The two motors chugged on southwards.

The surface was now hard wind-swept snow which was far better than sea-ice. But, in spite of this, the motor sledges only functioned for another three days. Day's car gave out first, a connecting-rod breaking through a piston. The big-end of the other went soon after.

Evans and his companions had now to get into harness and pull a sledge with such of the motors' load as they were able to take. It was exceedingly hard work but nobody regretted the change. The business of hauling the motors forward, fitting repairs, and stamping about in the cold while waiting for the engines to cool had not been pleasant. Even Day, who should have wept at the final breakdown, confessed that his disappointment was tempered with relief. And, after all, the motors had managed to take their heavy loads south for 51 miles. They had enabled the ponies to march light where the surface was worst. They had done what was required of them.

Evans organised the team so that the two tall men pulled in front with the shorter pair just behind them; the four of them were dragging a load of about 770 lb. They found that the worst hours of hauling were always the first and the last of each day. After a night's sleep they were stiff, then they would get tuned in and it was not until they began to feel tired towards the end of the day that once more they actively disliked it. Their orders were to wait for the others 62 miles beyond One Ton Depot, and, as they were determined not to be overtaken by the pony party and become a drag, they marched their fastest. Actually, in spite of their heavy load, they were out-distancing the main body every day, for Scott was deliberately keeping his marches down in order to give the weaker ponies a better chance.

The fortnight's sledge-hauling was uneventful. The diaries of the party over this period are typical of most sledging records. The going, rations (or the lack of them) and the temperature are almost the only things that matter on a polar journey. We read of the surface at first being 'rough but slippery', and then 'getting soft', which made their legs ache. Conditions are at their worst when the snow is covered with crystals and the runners do not slide smoothly. After one such march Lashly wrote: 'To-day it has been very hard work. The surface is very bad and we are pretty well full up, but not with food.'

They reached the rendezvous on 15 November, and had to wait there for six days before the main body arrived. It is improbable that any other party will ever spend a week's holiday on the Barrier. They built a giant cairn, and Day, 'gaunt and gay', entertained them and read *Pickwick Papers* aloud. The rest of the time was spent in sleeping, eating their frugal rations, and sleeping again.

* * * * *

The pony party set out on 31 October. Atkinson and Keohane led off with Jehu and James Pigg, two ponies which were considered to be poorer than the others and therefore in need of a day's start. Next morning the others completed their mail for the *Terra Nova*, which would arrive and depart again while most of them were away. Just before they left the base Scott asked one of the scientists what book he should take with him; he asked for something 'filling', and the scientist recommended a text-book on glaciology.

This idea did not appeal to Scott, and instead he took Browning.

There had originally been nineteen Siberian-Manchurian ponies, but of these only ten survived the voyage, the depot-laying journey, and the winter. Captain Lawrence Oates, cavalry officer and steeplechase rider – known, inevitably, as Titus after that first Titus Oates who so troubled Charles II – was in charge of them. Yet, strange to say, Oates never shared the great faith which Scott placed in the ponies. He was therefore in the unenviable position of knowing that the success or failure of the enterprise depended to a very large extent upon animals which he considered ill-chosen, but for which he had been made responsible. In his opinion the ponies were 'the most unsuitable scrap-heap crowd of unfit creatures'. Nevertheless he did all he could to make the ponies give of their best, and time and again Scott wrote in his diary words of admiration and gratitude for Oates's perseverance and devotion.

Wright's pony, Chinaman, was the first of the main-guard to be harnessed to his sledge, and Christopher the last. Chinaman was the slowest of all, but his amiable nature was always some consolation. Christopher, as usual, behaved like a demon. It took five minutes to throw him, and he was harnessed to his sledge while lying on the ground; but with a foreleg still tied up underneath his shoulder he managed to get to his feet and attempted to make off. Oates could never stop during a march, but had to go right through to the end of it, because once Christopher was stopped it was extremely difficult to start him again.

The first day's march, over the sea-ice from Cape Evans to Hut Point, showed the difference in speed of each pony. Before long they were strung out over several miles. Scott said that it reminded him of a regatta or a somewhat disorganised fleet with ships of unequal paces. That night he divided the party into three units according to their relative speeds. From then on Atkinson and Jehu always set out first, with Wright leading Chinaman, and Keohane leading James Pigg. Two hours later Scott, Wilson, and Cherry-Garrard followed with Snippets, Nobby, and Michael. Last of all came P.O. Evans – of *Discovery* fame – with Snatcher, Bowers with Victor, Oates with Christopher, and Crean with Bones. Snatcher was a wonderful little pony that always tried to take the lead, and Michael also stepped out well. With Christopher having to be taken through each march without stopping, this party always managed to catch up the first two before the end of the day. All three parties were self-contained units with their own tents, cookers and food-bags.

One by one they went over the tide-crack and up the slope on to the Barrier. They all instinctively paused at the top, not only to let man and beast get back their breath, but to take one last look round at the scene they knew so well. There cannot have been one among them who did not wonder when, and in what circumstances, he would see it again. Even Christopher was subdued.

The first stage of the Barrier journey was to One Ton Depot, and in order to avoid the crevasses they once again kept well to the east. The Barrier stretched ahead of them, silent, desolate, lifeless, the same today as yesterday and the day before, unchanging through an age of yesterdays and tomorrows – a monotonous flat shelf of floating ice over which ten eager men now pressed forward, leading ten grey ponies.

They took to travelling by night as it was better for the ponies, who could thus march during the coldest hours and rest in the warmth of the sun during the day, and, after a few days in which to get accustomed to the change, the animals slept and fed in far greater comfort than before.

At this time of year there is a considerable difference between midday and midnight. At both times the temperature may be many degrees below freezing-point, but in the daytime the rays of the sun are so hot that cracked and blistered faces are the rule rather than the exception. 'It is a sweltering day, the air breathless, the glare intense – one loses sight of the fact that the temperature is low [*minus* 22° F.]. One's mind seeks comparison in hot sunlit streets and scorching pavements, yet six hours ago my thumb was frostbitten.'[1]

1. All quotations are from Captain Scott's diary, unless otherwise stated.

Everybody was by now becoming accustomed to the tabloid feeding of sledging rations. At the beginning of the journey, even after a large plate of pemmican, they often felt hungry in the region of the throat though quite satisfied lower down. But after a few days the perpetual desire to chew something passed away to some extent, and they were content with a much smaller brew than at first. The biscuits were the only part of the rations that they could bite, and so always called for a measure of self-discipline. When each man drew his share of biscuits in the morning it was with the realisation that he would get no more that day, and so they had to be eaten with discretion. Everyone had a small linen bag in which he kept his sewing outfit, a diary, and perhaps a pocket edition of a favourite book; and little scraps of biscuit found their place amongst these other treasures.

The ponies were a continual source of anxiety. Although, during this stage of the journey, four different kinds of transport were used, motors, ponies, dogs and man-hauling, the ponies were the main feature and the others mere auxiliaries. The ponies were virtually to transport the base to the foot of the Beardmore Glacier, whence a picked team with two supporting parties would go forward to the assault. The concern for the ponies is evident in every page of their leaders' diaries, where any indications of improvement or collapse were noted and magnified. Snow walls had to be built to shelter the ponies whenever there was the slightest wind, and when the snow was drifting the men had to turn out time after time to dig them out; one or two of the restless animals would always kick the walls down, and rebuilding them added to the labour. However, on the whole the ponies did well at this time; they were pulling 700 lb. each except the two crocks, Chinaman and Jehu, who were only asked to pull 450 lb. The marches were kept down to 13 miles a day.

The two dog-teams, which had started later as they could travel faster, caught up the ponies on 7 November. Meares, who had brought the dogs from Siberia, had spent several years in that part of Asia, and had also made a remarkable journey into unknown Tibet. The other driver was Dimitri, a Russian boy, who had been engaged to help look after the dogs.

As is very often the case in the polar regions, particularly when the seasons are changing, conditions varied from day to day. On 8 November, Scott wrote in his diary: 'We are picking up last year's cairns with great ease, and all show up very distinctly. This is extremely satisfactory for the homeward march. With pony walls, camp sites and cairns, our track should be easily followed the whole way. Everyone is as fit as can be. It was wonderfully warm as we camped this morning at 11 o'clock; the wind has dropped completely and the sun shines gloriously. Men and ponies revel in such weather. One devoutly hopes for a good spell of it. . . .' Then two days later: 'A very horrid march. A strong head wind during the first part – then a snowstorm.'

On one of these days Wright stopped to look at the cyclometer on his sledge-wheel. Chinaman did not like being left behind and broke into a canter to catch up the others, whereupon, to everybody's astonishment, Jehu followed suit. The latter was such a poor specimen that he had scarcely been expected to reach the Barrier. From this moment Chinaman was christened the Thunderbolt, and Jehu the Barrier Wonder.

On 15 November they reached One Ton Depot, having covered 131 miles from Hut Point. After supper Scott called Oates and Bowers into his tent and a council of war was held. It was decided to rest for a day and then to push on at the rate of 15 miles a day, taking enough forage for the strongest ponies to reach the Beardmore Glacier. Some of them, clearly, could not go so far as that, and these would serve as food for the dogs. Scott's decision, with its implication that no ponies would be taken up the glacier, was heard with great relief. To lead a pony over the crevassed slopes near the foot of the Beardmore Glacier was considered nothing short of suicide. There was no ground for supposing that Scott had ever contemplated such a course, but nobody had liked to ask.

Lat. 80° 30′ S., where Evans's party was waiting, was reached without any great difficulty on 21 November. But the surface from One Ton Depot had been heavy and the ponies were tiring. Jehu was stopping frequently. 'It's touch and go whether we scrape up to the glacier; meanwhile we get along somehow.'

Both parties were glad to see each other. Evans thought the pony leaders looked tired; the latter noticed how thin the man-haulers had become. One of Evans's men facetiously remarked: 'We haven't seen anything of Amundsen.' Little did they imagine that he was at that moment four hundred miles ahead of them. They all camped together three miles farther on: sixteen men, ten ponies, twenty-two dogs, five tents and thirteen sledges.

They plodded on steadily with no landmark by which to mark their progress, and sometimes without any event to distinguish one day from another. Every morning they cooked their pemmican, struck camp and lashed up the sledge. Every evening they unlashed the sledge, put up the tent and ate their pemmican 'hoosh'. A simple business all of it, but with a detailed drill for each movement and calling for certain human qualities. And here a diversion will be justifiable in order to show what the stern reality of sledging entails.

The technique of sledging is more intricate than that of any other form of travel. The equipment is of a highly specialised order, and everything, from the tent to the snow-glasses, is the result of years of experience and long hours of careful thought and trial. The efficiency of every item is of supreme importance, since the penalty for an error leading to the breakdown of any article on a journey may be death by cold and hunger. So every detail must be scrupulously examined before starting out, and every eventuality guarded against. Nothing is too trivial for the gravest consideration, for in sledging all the little things count. A misfitting garment or the loss of a primus pricker might make just the difference between success and disaster. Under these conditions there is no room for anything but neatness: neatness of thought and neatness of execution. Sledging calls for particular qualities that do not necessarily show to great advantage in ordinary life, and tidiness and forethought are just as important as unfailing cheerfulness in the face of adversity. An untidy person is not only a nuisance but a very real danger.

Sledging is always a race. You take as much food as you can carry, principally pemmican, and then run as fast as you can, so that you get back to your base by the time that the food is finished. In order to save time on the journey it is of great importance to have a drill for all camping routine. Every party has its own particular arrangements, but they are all the same in principle.

Immediately the leader decides to halt for the night the business of camping begins. This is always chilly work, for a certain amount of moisture is bound to come from your hands and feet, and this freezes as soon as you stop moving. A wise man has a dry pair of inner gloves to put on at this time; but it is impossible to change your socks. First of all the sledge is unlashed, and then the ground-sheet is laid out, and the sleeping-bags, each one in its place, are laid upon it. Then the tent is lifted up and put down in position over the floor-cloth and bags. The cook for the week at once goes inside and starts to get dinner ready, while the others fill the cooking pots with snow and pass them in. After this everyone except the cook piles snow on to the tent skirting so that the tent cannot be blown down in a wind. If there are no observations to be taken or any other work to be done, everybody now crowds into the tent to watch the greasy brown lumps of pemmican and the greasy yellow lumps of butter being stirred into the hoosh. Pemmican is concentrated beef with extra fat added, and although it was invented more for the sustenance of the body than the comfort of the soul, there is no food like it for really hard work in really cold weather. The cook can vary the brew each meal by altering slightly the proportions of the ingredients. But he has not really very much opportunity for originality as all the possible and impossible mixtures are well known, and some people have a prejudice against chocolate in their pemmican and salt in their cocoa. Cooking a sledging supper sounds simple enough, but it must be remembered that there is very little room in the tent, especially when several men are changing their footwear or settling down simultaneously. Furthermore in spring and autumn the light is bad, and the cook's fingers may be covered with old frostbites and insensitive to anything except extreme heat and extreme cold. It is small wonder therefore that the precious hoosh is often spilt.

Before long the time comes when you are settled down inside the tent, with the pemmican just coming to the boil. When sledging in very low temperature this is about the only moment that is any good. It is a moment which, in its sudden soaring from the day's unpleasantness to heavenly comfort, almost makes up for everything else. Inside a sleeping-bag with a hot brew in front of you, while wind and driven snow whip the sides of the tent, it is almost possible to persuade yourself that you like sledging. Indeed, you can get a certain measure of enjoyment out of the multifarious tribulations of polar travel if you can acquire the right philosophy – the philosophy of the man who continually hit his head with a hammer because, he said, it was so nice when he stopped.

As soon as the meal is over the cook must collect the plates and spoons and keep them with the cooking pots. In a crowded tent a spoon has a remarkable habit of slipping away out of sight. The cook is not expected to clean them, for no dish-washing can be done on a sledging-journey. If a man is fussy about his tea tasting of cocoa he can stir a little water round the inside of his cup with a forefinger.

The great secret of avoiding frostbite is to dry gloves and footwear thoroughly each night. Hands and feet are so far from the heart that the circulation is liable to fail there earlier than elsewhere, and in most people they produce an astonishing amount of sweat. So every man except the unfortunate cook changes his socks as soon as he gets inside the tent, and takes immediate steps to dry them and his gloves. Several pairs are always worn, and hoar-frost forms between each layer so that each one sticks to the next and has to be peeled off. After being thoroughly scraped with a spoon they are hung up in the peak of the tent. But this does not by any means end your responsibility, for now, after supper, you cannot go to sleep leaving it to Providence to do the rest; you must be continually turning these garments inside out, and then the dampest must be put inside your sleeping-bag next to your body. Having thus disposed of your gloves and footwear you settle down to a very short smoke, to write up your diary and perhaps to work out a computation. However uncomfortable you may be, sleep comes readily enough and the morning all too soon.

So much for the routine of sledging. We must get back, after this long digression, to Scott's party, now half-way between the Upper Barrier Depot and the Middle Barrier Depot. The ponies were daily getting thinner, but they were still pulling well, even the crocks sticking gamely to their work. Jehu was the weakest, and Meares and Dimitri were looking eagerly for the chance to feed their dogs on him, but Atkinson and Oates were anxious to get him beyond the point where Shackleton had killed his first pony, 'Chinaman'. On 24 November Jehu was led back along the track and despatched with a bullet through his brain. He had gone 15 miles farther than 'Chinaman', but this can scarcely have given him any joy. Lieutenant Evans's team were particularly grateful for the pony-meat cooked in their pemmican hoosh. They had been in harness for some time now and were considerably more hungry than the others, who found the ration sufficient for the less strenuous exercise of leading ponies or driving dogs. The fresh meat, with its antiscorbutic properties, did them all far more good than was realised in those days when nobody had ever heard of vitamin 'C'. Day and Hooper turned back that day, and in due course reached Cape Evans.

They were now only 150 miles from the Beardmore Glacier, but Scott was doubtful if they could reach it, much less the Pole. 'It is still rather touch and go. If one or more ponies were to go rapidly downhill we might be in queer street.' They were having to fight for each day's march, and when poor old Chinaman, who should have been a pensioner, was killed, Cherry-Garrard commented that he was well out of it. The surface was very heavy for pulling, while dull weather strained their eyes and made steering difficult. Bowers wrote: 'We have now run down a whole degree of latitude [69 miles] without a fine day or anything but clouds, mist and snow from the south.'

Gone for the moment were the few pleasures of the

march. No more enjoyable luncheons at half-time, with occasional interesting talks about books and flowers and pictures, memories of the past. During those days of toil talk was an unnecessary waste of valuable time and breath. 'Pack up', 'All ready?', 'Spell Ho!', 'Camp Ho!', 'Wind watches', was all the conversation necessary. At the end of each march every man was worn out.

But better things were in store. Groping, as they were, through blizzards and mist, they were quite unprepared for the sight that suddenly greeted them on 29 November, as they were passing the landmark of Scott's Farthest South, 82° 17′. The clouds rolled back, and towering above them was the great rampart of mountains which would soon cross their path to the South. The triple peaks of Mount Markham seemed to be almost on top of them. After 300 miles of the bleak and desolate Barrier this was a most exhilarating and encouraging sight. That night, after 29 days' travelling, they camped four miles beyond the point where Scott had turned back in 1902. On that occasion he had covered the distance in 58 days, while Shackleton, six years later, had taken 23 days.

The next four days were mostly fine, and everyone was in excellent spirits. They might well have been inspired by such magnificent scenery. Although they still referred to the day and the night, there was by now no great difference between the two since the sun merely travelled round the heavens. At midnight its altitude was about 12 or 13 degrees and at noon 28 or 29. This meant that there were always interesting light and cloud effects, which enhanced the beauty of the scenery a thousandfold. They started sledging each evening in a wonderful sunset, and as the sunset slowly flooded into sunrise the mountains changed before their eyes from brown to crimson and then to gold.

On 1 December Oates took Cherry-Garrard's place in Scott's tent. The polar party was beginning to take shape. Cherry-Garrard had never expected to be a member of it, for Scott had expressly stated that in his opinion the younger men did not stand the cold so well as the older ones, and had mentioned Cherry-Garrard, who was 26, as an example. At the beginning of the journey Scott and Wilson were the only certainties. Dr Wilson appears to have been one of the most lovable men on earth. Scott wrote of him: 'Words must always fail me when I talk of Bill Wilson. I believe he really is the finest character I ever met – the closer one gets to him the more there is to admire. Every quality is so solid and dependable. . . . Whatever the matter one knows Bill will be sound, shrewdly practical, intensely loyal and quite unselfish.' Whoever else was left behind, Wilson, Scott's dearly-loved friend and tried sledging companion of the old *Discovery* days, would certainly be taken.

That night the pony Victor, 'as gentle as a dear old sheep', was shot, for pony-food was running short. Little 'Birdie' Bowers was very upset. He had grown inordinately fond of him and always managed to feed him with a biscuit out of his daily ration. He gave him the last just before the poor beast was led away to execution. That night he wrote: 'Victor did a splendid march and kept ahead all day, and as usual marched into camp first, pulling over 450 lb. easily. It seemed an awful pity to have to shoot a great strong animal. . . . Good old Victor! . . . He has done his share in our undertaking, anyhow, and may I do my share as well when I get into harness myself.' The harsh school of sledging makes brutes of most of us, but Scott was always affected and sometimes wept at these butcheries of dogs or ponies.

They travelled 23 miles in the next two days and then camped opposite the Gateway, as Shackleton had named the place where he passed between Mount Hope and the mainland. From here they could see the outlet of the Beardmore Glacier, flowing in waves of snow-covered pressure-ice. In front of them was the glory of mountains, range upon range, peak after snow-clad peak. They went into their tents that night excited beyond measure. The Barrier Stage was almost over, and Scott hoped to sleep upon the glacier the next night. 'We should reach it easily enough on to-morrow's march. . . .' Tomorrow, tomorrow . . .

But it was not to be. Far away, up on the plateau above them, a little cold wind awoke. Blowing the snow off the peaks and the passes, down the valleys and slopes, it came tearing and plunging. Growing in strength and in noise and in volume it swept down the glaciers and on to the Barrier. A raging howling blizzard struck the camp. It lasted *four days*.

This was a most cruel blow. They were a day's march from the Beardmore: fourteen men, ponies and dogs, all in good health and high spirits, and ready to make a great advance. And now there was nothing to do but sleep and eat, and to eat food that should have taken them many miles farther, for it was impossible to reduce rations so early in the journey.

The tents were soon snowed up, and the weight of snow pressed the canvas inwards, so that the men who slept next to the walls were cramped for space. The weather became warmer and the snow changed to driving sleet. But the blizzard continued without a pause.

The wretched ponies suffered most. Drift-snow, which found its way under their rugs, melted and made their coats wet, and the water dripping to the ground drained all the heat from their bodies. They could only stand and shiver with hanging heads and bent knees and the wild frightened eyes of animals in pain. But the dogs were quite happy lying curled up in balls, quite oblivious of the snow that swirled and drifted over their heads.

And still the blizzard went on.

The temperature rose above freezing-point, and the surface became 18 inches deep in slush. Bowers wrote: 'We are wet through, our tents are wet, our bags, which are our life to us and the objects of our greatest care, are wet; the poor ponies are soaked and shivering far more than they would be ordinarily in a temperature of fifty degrees lower. Our sledges – the parts that are dug out – are wet, our food is wet, everything on and around and about us is the same – wet as ourselves and our cold, clammy clothes. Water trickles down the tent poles and only forms icicles in contact with the snow floor. The warmth of our bodies has formed a snow bath in the floor for each of us to lie in. This is a nice little catchwater for stray streams to run into before they freeze. This they cannot do while the warm human lies there, so they remain liquid and the accommodating bag mops them up. The rest of the show is indescribable. I feel most for the unfortunate animals and am thankful that poor old Victor is spared this. . . . This idleness when one is simply jumping to go on is bad enough for most, but must be worse for Captain Scott. I feel glad that he has Dr Bill in his tent; there is something always so reassuring about Bill, he comes out best in adversity.'

During the blizzard Oates suffered more than the other men, for most of the time he was looking after the ponies. As he brought more wet in with him whenever he re-entered the tent, he took to spending the greater part of the day outside, crouching under the lee of one of the pony walls.

On the fourth day the wind began to drop about noon, and Scott went out two or three times that night to look at the weather. The party was up at 5.30 a.m. and away at 8. As was to be expected, the surface was appalling; the ponies sank to their bellies and lay down until driven to their feet again; the dogs were almost swimming. All day they fought their way forward, not daring to stop for lunch for they knew they would never get started again. At last, after nearly fifteen hours' continuous struggling, they camped, utterly exhausted.

That the five remaining ponies had brought them to the threshold of the Beardmore Glacier was largely due to Oates's unremitting care. Wilson now turned to him and said: 'I congratulate you, Titus,' and Scott, in his quiet voice, added: 'And I thank you, Titus.'

One by one the ponies were led back along the track, and there, under the shadow of Mount Hope, they were slaughtered. The meat was cut up for food, and the dogs scattered the entrails over the scarlet snow.

They called it Shambles Camp.

TWO

The Beardmore Glacier

(126 miles: 10–21 Dec.)

First Sledge	*Second Sledge*	*Third Sledge*
Scott	Lieut. Evans	Bowers
Wilson	Atkinson	Cherry-Garrard
Oates	Wright	Crean
P.O. Evans	Lashly	Keohane

THE BARRIER Stage had been completed more or less according to schedule. The ponies had taken twenty-four units of food to within five miles of the objective. But the party was five days late owing to the blizzard, an occurrence which nobody could have foreseen in December. This meant that they were already eating rations that should not have been touched until they were on their way up the glacier. The Pole was still 500 miles away, and the season slipping by. In eleven days it would be Midsummer's Day.

But there was an even more serious loss which could not be reckoned by calendar or diet-sheet. The party, having completed a strenuous 400-mile journey, had of necessity lost a proportion of its health and vigour. This became only too evident after the first day of hauling sledges up the glacier.

It was noon before the necessary reorganisation had been completed and the parties were able to get under way. The surface was soft, and when the slope became steeper skis had to be taken off. The pulling was now very tiring, for with every stride the men sank deep into the snow. Lieutenant Evans's team came in an hour after the others, and that night Scott commented in his diary: 'I have not felt satisfied about this party. The finish of the march today showed clearly that something was wrong. They fell a long way behind, had to take off ski, and took nearly half an hour to come a few hundred yards. True, the surface was awful and growing worse every moment. It is a very serious business if the men are going to crock up. As for myself, I never felt fitter and my party can easily hold its own. P.O. Evans, of course, is a tower of strength, but Oates and Wilson are doing splendidly also.' Lieutenant Evans and Lashly had now been in harness for five weeks, and Wright and Atkinson since the death of their ponies, so these four men were not so fit as the others who had that day put on harness for the first time. Scott was worried.

They camped that night in a strong wind which they did not regret, for they knew that the wind would blow the loose powdery snow right across the

Barrier and into the sea. Every hour of wind would make the surface harder.

Next morning, before starting, they provisioned the Lower Glacier Depot with enough food to take the three returning parties back to the Lower Barrier Depot. The Lower Glacier Depot was the fourth cache south of One Ton Depot. These depots were on an average 69 miles apart and contained seven days' food for each unit of four men. As the distance was too great for provisions to be carried all the way, this depot system was unavoidable, and for a return journey over the outward route there could be no objection to it provided it were only possible to ensure that every party could get to the next cache. Nothing can save a man-hauling party that is unable to reach or find any one depot.

The dog-teams turned back that day. They had been taken about 160 miles farther south than was originally intended, and in order to provide for this extension everybody sacrificed a biscuit a day going up the Beardmore, and Meares and Dimitri went short of one meal a day all the way back to Cape Evans.

After the dogs had left it was an anxious moment when the men began to haul, for they wondered whether they could pull the extra load. The snow was still very deep, and periodically the sledges sank to the cross-bars, when only a truly awful struggle would restart them. The starting was worse than pulling as it required from ten to fifteen desperate jerks on the

1. Captain Scott

2a. The *Terra Nova* in pack-ice

2b. Looking across the pack-ice from the main-top

3a. Seals basking on pancake ice off Cape Evans

3b. Weddell seal with her calf

4. The *Terra Nova* from a grotto in an iceberg

5. Close by an iceberg

6a. Shackleton's hut under Mount Erebus

6b. Captain Oates and the ponies think things over

7a. Dr. Atkinson's dog-team landing stores from the ship

7b. Picketing ponies

8a. Day and Lashly fixing one of the motors

8b. By contrast, Sno-cat Able used in the 1958 Trans-Antarctic Expedition

9a. Bowers, Wilson and Cherry-Garrard ready to start for Cape Crozier

9b. The first hot meal on their return

10a. The end of Barne Glacier

b. A sketch by Dr. Wilson showing sledging in typical Antarctic conditions

11a. Dr. Wilson sketching on the Beardmore Glacier

11b. Another dramatic sketch by Dr. Wilson

12a. Captain Scott writing his diary

12b. Captain Scott's birthday party

13a. Sledging party at Three Degree Depot

13b. Camp at Three Degree Depot

14a. and b. On the march to the Pole

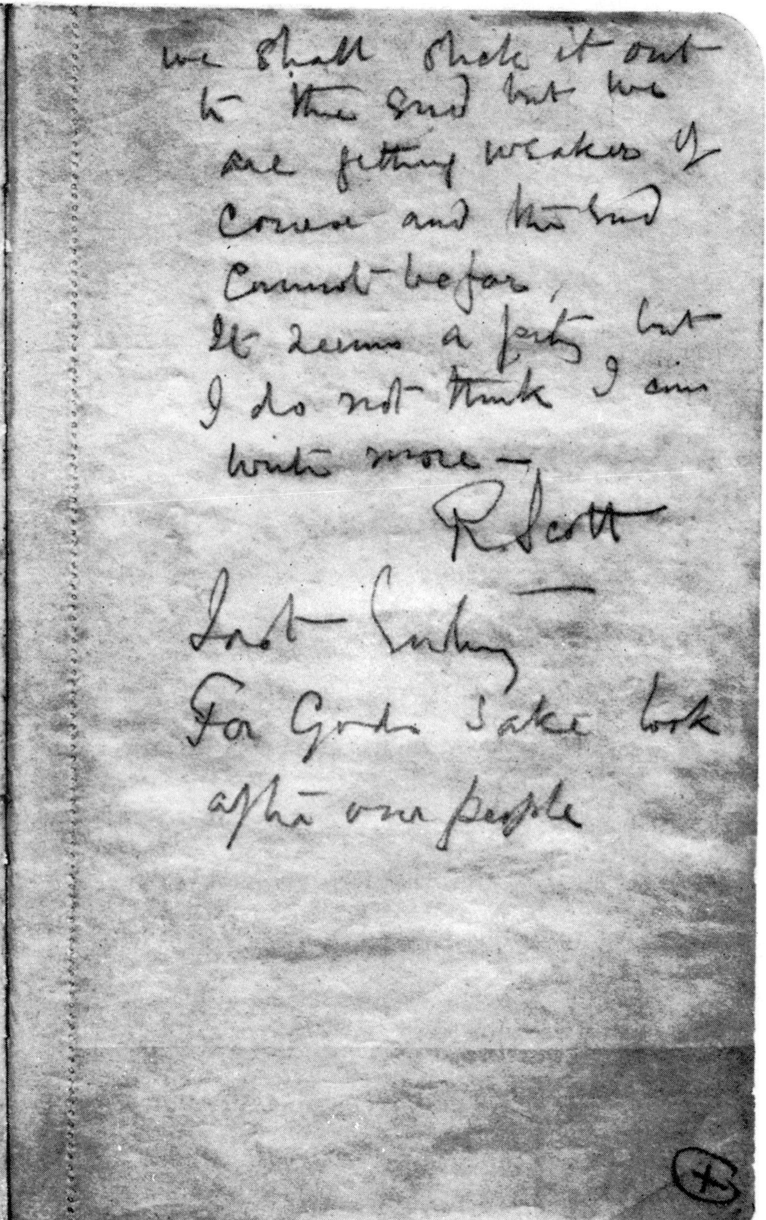

we shall stick it out
to the end but we
are getting weaker of
course and the end
cannot be far,
It seems a pity, but
I do not think I can
write more —
R. Scott —
Last entry
For God's sake look
after our people

15. The last page of Scott's diary

16. The five at the Pole

Oates Bowers Scott Wilson Evans

harness to move the sledge at all. '. . . We stuck ten yards from the camp, and nine hours later found us little more than half a mile on. I have never seen a sledge sink so. I have never pulled so hard, or so nearly crushed my inside into my backbone by the everlasting jerking with all my strength on the canvas band round my unfortunate tummy. We were all in the same boat however.'[1]

There can be little doubt that the first three days from the Lower Glacier Depot was the hardest physical test of the outward journey. In addition to all the difficulties that they had encountered in soft snow on the Barrier below, they were now pulling uphill, and with the load increased to 800 lb. a sledge – that is to say, 200 lb. a man – from the moment the dog-teams had left. The sledges, which were so overloaded that they often capsized, left tracks like snow-ploughs behind them. Every man was stripped down to his vest and still bathed in perspiration; the perspiration fogged their snow-glasses, and when they removed them they experienced the agonies of snow-blindness. Half the party were soon suffering from it. 'I have missed my journal for four days, having been enduring the pains of hell with my eyes as well as doing the most backbreaking work I have ever come up against.'[2] After nine hours of almost superhuman pulling each day they camped, utterly worn out and a bare four

1. Bowers.
2. Bowers.

miles farther on. But the beauty of the glacier, with its light blue patches of ice breaking through the whiteness of the snow flowing between the steep dark granite cliffs of the surrounding mountains, was a great compensation for the hard work. Nobody regretted the easier but exceedingly monotonous travelling across the Barrier.

There is no doubt that Scott's team, consisting of the picked men, was the strongest of the three. It was in front all the way and never had any difficulty in keeping its place. This, of course, was as it should have been, for Scott had to lead the way up the glacier, which he did with excellent judgement. On 19 December Bowers wrote: 'There is no doubt that Scott knows where to aim for in a glacier, as it was just here that Shackleton had two or three of his worst days' work, in such a maze of crevasses that he said that often a slip meant death for the whole party. He avoids the sides of the glacier and goes nowhere near the snow; he often heads straight for apparent chaos, and somehow, when we appear to have reached a *cul-de-sac*, we find it an open road.' And later: 'Scott is quite wonderful in his selections of route, as we have escaped excessive dangers and difficulties all along.'

The Beardmore Glacier is so vast that it dwarfs its surroundings, and the big tributary glaciers that feed it seem almost insignificant. Fossilised plants and ferns show that, perhaps many millions of years ago, these mountains were once covered with thick vegetation.

On 14 December things began to go a little better, and the party made about 9 miles. The going was improving, the snow being more closely packed and therefore suitable for skis. But, with the absence of the deep snow, crevasses began to appear, and when they were camping that night Crean put his foot into one which ran in front of his tent door. They dropped an empty oil-can down it, and heard it banging from side to side for a 'terribly long time'. The aneroids gave their height as 2000 feet above sea-level. They were climbing at a rate of about 500 feet a day.

From now onwards weather and surface were favourable; the deep snow was left far behind, and there was nothing to stop their progress except the limitations of the human body, the necessity for rest and sleep. It was warm sunny weather, and often when they halted Wilson sat on the sledge and sketched. These sketches, apart from being full of artistic merit, were of great assistance to the survey. Evans and Bowers plotted the mountains, and names were given to the features that Shackleton had not already christened. They had Shackleton's notes and diaries with them, which were a considerable help, and their course from One Ton Depot had been set by his chart. All through the journey they discussed his march and compared it with their own. And now they were six days behind him, and hurrying for all they were worth.

The most conspicuous mountain Shackleton had

named the 'Cloudmaker' because a little cloud always clung to the peak. He gave its height as 9971 feet, and the odd 1 foot is a pretty compliment to Marshall, his cartographer, in view of the difficulty of determining heights on a journey of that nature. The three secondary ranges south-west of the Cloudmaker he called the Adams, Marshall, and Wild Mountains, after his three companions.

Sunday, 17 December, proved an exciting day. They got into big pressure waves with blue ice on the crests and deep snow in the hollows. Pulling the sledges up to the top of each wave was hard work, but going down the slope on the far side was quite a different matter. All they could do then was to set the sledge straight and hold tight. The wind whistled in their ears as they shot down to the bottom, and their impetus carried them some way up the other side. Cherry-Garrard said it reminded him of the scenic railway at Earls Court. 'For once we can say, "Sufficient for the day is the good thereof." Our luck may be on the turn – I think we deserve it. In spite of the hard work everyone is very fit and very cheerful, feeling well-fed and eager for more toil.'

Before starting off on 18 December they laid the Middle Glacier Depot, in Lat. 84° 36′ S. The Barrier depots had all been marked with black flags, but this time a red flag was used to show against both the ice and the dark cliffs. Unfortunately no bearings had been taken the previous night, and when they broke camp it was snowing; in consequence all three returning parties had difficulty in finding this depot.

As they climbed higher and higher each day it became noticeably a little colder. Their sunburnt hands and faces began to smart and their lips were very sore, which caused a run on the medical stores. They also had one or two bad bruises from falls into crevasses, though nobody had yet gone down to the full length of his harness. On the blue ice crevasses were quite conspicuous, being either open or transparently bridged over, and it was advisable to stride well over any line of snow. Nevertheless, in harness this was not very easy, as the trace would sometimes jerk the man back in mid-air just as he was about to step over, and the next moment he would be struggling to get clear without stopping the sledge.

20 December was another eventful day. The going was even better, and they soon realised that they were travelling at a speed of two miles an hour. Scott, who had come to the end of his diary as he had been writing on one side of the paper only, had the night before reversed the book and started to write on the other side of each page. He had hoped that this was also turning over a new leaf in the fortunes of the Expedition, and he felt that it was already bringing them luck.

At the lunch halt Bowers discovered that the cyclometer had dropped off his sledge-wheel. This was a disastrous loss, since it meant that one of the three returning parties would have no means of knowing the distance travelled between astronomical observations. Navigation would be much more difficult, and navigation on the Barrier or the plateau, where finding a depot is like picking up a buoy in the middle of an ocean, is a matter of life and death. Bowers was very much upset. Christopher, in one of his vicious moods, had originally kicked off the sledge-meter, but Bowers had lashed it on again with raw hide thongs. It had held until just recently, when he had noticed that it was working loose again and had sat up half the night repairing it; but since then a screw had been dislodged when the sledge bumped over rough ice, and the gear-wheels had fallen out of their case. Bowers and Wilson walked back along the track for two miles but saw no sign of it.

Pulling 160 lb. a man, they did almost 23 miles that day, by far their best glacier march, and camped 6500 feet above sea-level. They were now less than three days behind Shackleton's time.

That night Scott told them whom he had chosen to go on with him over the plateau, and who would return after one more march. Cherry-Garrard wrote in his diary: 'This evening has been rather a shock. As I was getting my finnesko on to the top of my ski beyond the tent Scott came up to me and said that he was afraid he had rather a blow for me. Of course, I knew what he was going to say, but could hardly grasp that I was going back – tomorrow night. The returning party is to be Atch, Silas, Keohane and self.

'Scott was very put about, said he had been thinking a lot about it, but had come to the conclusion that the seamen, with their special knowledge, would be needed: to rebuild the sledge, I suppose. Wilson told me it was a toss-up whether Titus or I should go on; that being so, I think Titus will help him more than I can. I said all I could think of – he seemed so cut up about it, saying, "I think, somehow, it is specially hard on you." I said I hoped I had not disappointed him, and he caught hold of me and said, "No – No – No," so if that is the case all is well. He told me that at the bottom of the glacier he was hardly expecting to go on himself; I don't know what the trouble is, but his foot is troubling him, and also, I think, indigestion.'

Scott's entry is: 'I have just told off the people to return tomorrow night: Atkinson, Wright, Cherry-Garrard and Keohane. All are disappointed – poor Wright rather bitterly, I fear. I dreaded the necessity of choosing – nothing could be more heartrending. I calculated our programme to start from 85° 10′ with 12 units of food and eight men. We ought to be in this position tomorrow night, less one day's food. After all our harassing trouble one cannot but be satisfied with such a prospect.'

Atkinson was to be in charge of the party, Wright being the navigator. Lieutenant Evans sat up very late that night preparing them a copy of the route.

Scott was now aiming to pass between the Dominion

Range and a nunatak which Shackleton had named Buckley Island. The congestion of ice between these two and at the junction of the glacier and the plateau caused a mass of pressure-ice which had to be crossed. Scott had originally intended to pass to the west of Buckley Island, but the wild tangle of pinnacles and crags on that side looked even more formidable than on the other. They made, therefore, for a slope near the end of the nunatak, which was presumably Shackleton's place of ascent as there seemed to be no other route. They did not, however, go as near to the land as he had done and so avoided some of the difficulties which he had encountered. All the way up the glacier Scott's policy had been to keep well out in the middle, and by this means he had reduced Shackleton's lead from six days to two. 'Scott is quite wonderful in his selections of route . . .'[1] However, when they reached the top of the slope they found themselves in bad pressure-ice, with no possible way round. Where the crevasses were too wide to jump over there was nothing for it but to step on the snow bridge and hope for the best. Scott, being in the lead, always had the first chance of falling through, but very often the first men to step on the bridge weakened it, and those who followed fell through. In the course of the morning every man went down a crevasse, sometimes to the full length of the harness. 'We had to go over dozens by hopping right on to the bridge and then

1. Bowers.

on to the ice. It is a bit of a jar when it gives way under you, but the friendly harness is made to trust one's life to. The Lord only knows how deep these vast chasms go down, they seem to extend into blue-black nothingness thousands of feet below.'[1]

Passing Buckley Island they saw among sandstone and old granite the stratification which Shackleton had found to be *coal*, a discovery of the greatest importance, as it proved that this ice-bound land was once covered with dense forest.

Meanwhile the wind was veering, and by noon it was blowing straight up the glacier, bringing a mist with it, and just when they were in the worst pressure-ice they had to stop and camp. While they were waiting for the weather to clear, Bowers checked the list of rations that were to be left at the Upper Glacier Depot, and Scott wrote letters which were to be taken back from there. In one of them he said: 'We are struggling on, considering all things, against odds. The weather is a constant anxiety, otherwise arrangements are working exactly as planned.'

At 3 o'clock the fog lifted and they were able to continue their march. They soon got beyond the crevasses and started climbing a series of ridges, but now that new snow had replaced glacier ice the pulling was much harder. Scott seemed to be wound up and went on and on. Every rise they climbed fired him with a desire to climb the next; and every rise had

1. Bowers.

another beyond and above it. They camped at 8 p.m., and as they had covered 11 miles with a rise of 1500 feet, they were all pretty weary.

It was a sad camp that night, in spite of an extra large pemmican hoosh to celebrate another milestone safely reached. Wilson went into Cherry-Garrard's tent to say good-bye; he said that he fully expected to be sent back with the next party as he could see that Scott intended to take on the strongest, perhaps the three seamen. Cherry-Garrard distributed all his spare kit and the last of his precious tobacco. There were fur boots and a piece of handkerchief for little Bowers, pyjama trousers for Wilson, half a scarf for Crean, and tobacco for Scott and Oates. Bowers worked till very late, making the allotment for the returning party and dividing the weights between the two sledges which were to go forward, while Scott wrote: 'For me it is an immense relief to have the indefatigable little Bowers to see to all detail arrangements of this sort.'

And in a letter: '. . . we are practically on the summit, and up to date in the provision line. We ought to get through.' Through to the Pole this would be; for already, I think, he was doubtful of the ultimate issue. You must read his diary and judge for yourself. There seems to be no joy in his writing at this time; no glorious exultation at the prospect of reaching his goal. '*We are struggling on, considering all things, against odds. The weather is a constant anxiety . . .*'

THREE

The Plateau

(359 miles: 22 Dec.—16 Jan.)

First Sledge	*Second Sledge*
Scott	Lieut. Evans
Wilson	Bowers
Oates	Lashly
P.O. Evans	Crean

THEY HAD reached the top of the Beardmore Glacier on Midsummer's Day, and that night the temperature went down to zero. As it had not been so low for some time, this seemed to be a warning that the summer was in decline. They ought to have been camping at the Pole that night, ready to begin the return journey on the first day of the second half of the season. But the Pole was still 372 miles away.

After breakfast on 22 December they built the Upper Glacier Depot. Ice-axes, crampons, and all the glacier equipment were left there, together with as much personal gear as they could dispense with, and 3½ days' food to take the eight advancing men back, when the time came, to the Middle Glacier Depot. Most of them were rather moved at saying good-bye. Bowers wrote that Cherry, Atch, and Silas were quite overcome, and Scott refers to the returning party as 'dear good fellows'.

Once more, but for the last time, the weights on the outgoing sledges were increased to a full load. From now onwards the burden would every day get lighter. Scott led off at a brisk pace, but the other sledge kept close up, which showed that its team had been well chosen. Looking back along their tracks they could see the depot cairn with its black flag for more than three miles, which promised well for finding it again. The first day was not a long one, for they started late after making the depot and were afterwards delayed by having to mend the sledge-wheel. In 7 hours they covered 12 miles.

For the first two days Scott steered south-west as there were huge pressure ridges in every other direction. This course was scarcely satisfactory, but they were climbing the whole time, which was a hopeful sign. The crevasses they crossed were wide enough to have swallowed up a battleship, let alone a sledging party. They rushed the sledges over these places and only occasionally put a foot through at the edge. It was noticed that whenever a snow bridge was rotten it was now at the lower edge, whereas on the glacier it had always been at the upper edge.

Towards the end of the second day the horizon

levelled in every direction. It was the beginning of better things. When they camped that night they had covered over 17 miles in 8½ hours and had climbed 800 feet. They felt confident that they had at last reached the summit proper, and everybody was very cheerful at having said good-bye – as they thought – to steep slopes with their attendant crevasses. Scott wrote: 'To me for the first time our goal seems really in sight. We can pull our loads and pull them much faster and farther than I expected in my most hopeful moments. I only pray for a fair share of good weather. There is a cold wind now as expected, but with good clothes and well fed as we are, we can stick a lot worse than we are getting. I trust this may prove the turning-point in our fortunes for which we have waited so patiently.'

On Christmas Eve they returned to their proper course of true South and travelled just over 16 miles. But they were not yet beyond the pressure ridges, as Scott had hoped, and another four days elapsed before they passed the last crevasses.

Lashly celebrated Christmas Day, which was also his forty-fourth birthday, by falling down a crevasse to the full length of his harness and trace. It was indeed fortunate that Bowers had noticed a few days previously that Lashly's rope was becoming worn and had replaced it with a new one. As it was Christmas Day Bowers produced a stick of chocolate and two spoonfuls of raisins apiece for lunch, in addition to the usual ration. They had an extra long halt that day and let the primus stove burn on long after the tea had been made. The smokers lit their pipes and the talk was of Christmas. How proud Scott must have felt of his companions as they sat round him chatting away brightly, making no reference to the difficulties of the journey and the discomforts involved. They were full of good cheer, with the nicest thoughts for each other and for those at home, of whom Wilson unconsciously reminded them by humming: 'Keep our loved ones, now far absent, 'neath Thy care.' All through the day they were sustained by thoughts of the vastly different Christmases to come to them later. But for Lashly and Crean alone did dreams come true, for Teddy Evans, the only other survivor, was to lose his wife as well as his five companions.

The afternoon march was oppressively warm as the breeze had died down. Perspiration fogged their glasses, and everybody longed for the final halt which seemed so long in coming. At last Scott stopped, and they looked at the sledge-wheel which registered 14½ (geographical) miles. Scott said: 'How about 15 miles for Christmas Day?' and on they gladly went to complete the distance. That evening they celebrated in the time-honoured manner of over-eating, for Bowers had kept quite a banquet hidden out of the official weights. So '. . . we had a pretty good tuck-in. Dinner consisted of pemmican, biscuits, chocolate éclair, pony-meat, plum-pudding and crystallised ginger and four caramels each. We none of us could hardly move.'[1]

They experienced the almost bewildering sense of happiness that comes to a half-starving man with his first good meal. After the others were asleep Bowers turned to Evans and said: 'Teddy, if all is well next Christmas we will get hold of all the poor children we can and just stuff them full of nice things, won't we?'

The march over the plateau continued without event. The temperatures were fairly consistent, being about *minus* 3° F. by day and *minus* 10° F. by night. But ceaselessly from the South blew what Shackleton had called the 'pitiless, increasing wind'. This kept them cool while pulling and, if it had not been for the fact that their faces so frequently became ice-covered, it would at this stage have been pleasant. But before they were warmed up at the beginning of each march, and when camping at night, they felt the cold acutely. They were always tired out by the end of the day.

On 27 December Bowers wrote: 'There is something the matter with our sledge or our team, as we have an awful slog to keep up with the others. I asked Dr Bill and he said their sledge ran very evenly. Ours is nothing but a desperate drag with constant rallies to keep up. We certainly manage to do so, but I am sure we cannot keep this up for long. We are all pretty well done up tonight after doing 13.3 miles. Our

1. Lashly.

salvation is on the summits of the ridges, where hard névé and sastrugi obtain, and we skip over this slippery stuff and make up lost ground easily. In soft snow the other team draw steadily ahead, and it is fairly heart-breaking to know you are pulling your life out hour after hour while they go along with little apparent effort.'

Next day the two teams changed over and found conclusively that Evans's sledge was the cause of the trouble; it had been distorted by bad strapping and bad loading. Oates had nearly lost his sleeping-bag a few days earlier through not fastening it on tight enough, and in order to avoid such a fatal calamity the second party had gone to the other extreme and warped their sledge with too tight strapping. After supper Scott went into the other tent and told them plainly that they must wrestle with the trouble and put it right.

29 and 30 December were both tiring days as the surface had deteriorated. The drift-snow carried by the eternal southerly wind had accumulated in patches which made the going heavy. Their distances dropped to 14 and 13 miles respectively. On the 29th Scott wrote: 'The satisfactory thing is that the second party now keeps up, as the faults have been discovered . . .' but the following day it lagged behind again and camped three-quarters of an hour after the others. There is no doubt that this team was tiring rapidly. Lieutenant Evans and Lashly had already been in

harness for nearly 700 miles and the strain was beginning to tell. That night Scott wrote: 'We have caught up Shackleton's dates. Everything would be cheerful if I could persuade myself that the second party were quite fit to go forward.' While Evans's entry was: '. . . too tired to write up diaries even.'

Before starting out on 31 December Scott told the second team to leave its skis and some other equipment, weighing in all about 100 lb. This was equivalent to saying that he intended to take his own team, as it was then constituted, to the Pole, for it was almost certain by this time that he *would* reach the Pole. They only marched until noon that day as Scott had decided to strip off the worn 12-foot sledge-runners which had been cut and scratched coming up the glacier, and to substitute new 10-foot runners that had been brought for the purpose. This job was done during the afternoon and evening by the three seamen. It took them until 11 p.m., and then they stayed up to see the New Year in. Meanwhile the others built a cairn containing seven days' food to take each of the two teams back to the Upper Glacier Depot; this cache was called Three Degree Depot, being almost at Lat. 87° S., and three degrees from the Pole.

During the next three days they continued to climb, and the thermometer fell steadily. On New Year's Day they camped at 9600 feet with the temperature *minus* 14° F., and on 3 January at 10,180 feet with it *minus* 18° F. The magnetic variation was 180°, showing that they were between the Magnetic Pole and the Geographical Pole, and in order to go South they had from now on to march north by compass. On 2 January they were visited, to their astonishment, by a skua-gull which appeared from the South. It settled in front of them several times, but fluttered off when they tried to catch it.

On the morning of 3 January, at Camp 55, Scott made the fateful announcement. He would go ahead with Wilson, Oates, P.O. Evans and Bowers. Lieutenant Evans would return next day with Lashly and Crean. The die was cast.

He wrote in his diary that evening: 'Within 150 miles of our goal. Last night I decided to reorganise, and this morning told off Teddy Evans, Lashly, and Crean to return. They are disappointed, but take it well. Bowers is to come into our tent, and we proceed as a five-man unit tomorrow. We have 5½ units of food – practically over a month's allowance for five people – and it ought to see us through.'

This is the bald record of the most important decision in Scott's career. It gives no hint of the days and weeks of labour striving to decide, and the agony of deliverance when he made his intention known. For Scott was one of those hypersensitive beings to whom inflicting hurt on others was to himself a tearing physical pain, and he knew that to Teddy Evans, Lashly, and Crean this announcement, although not unexpected, would be the cruellest of disappointments. Therein lies the reason for his deciding, at the last moment, to take on an extra man. On 31 December he left four pairs of skis behind, for he intended that only four men should go to the Pole. But in the next four days he changed his mind and decided to take one man more. This was Bowers, who deserved, if any man did, his place in the polar party.

Little 'Birdie' Bowers was a quaint but capable person with a capacity for work that was prodigious. There was only 5 feet 4 inches of him, but all of it was pure gold. He had helped Scott considerably when in charge of the stores on the ship, in the routine work at Cape Evans, and in preparing weights and measures for the southern journey. 'For me it is an immense relief to have the indefatigable little Bowers to see to all detail arrangements of this sort,' and later, when he had been more severely tested, Scott wrote: 'I believe he is the hardest traveller that ever undertook a Polar Journey as well as one of the most undaunted.'

All the organisation for sledge travelling was for the four-man unit: the rations were made up for four men for one week; the tents held four men and the cookers four mugs, four plates and four spoons. Now, at the turning-point, there were only four pairs of skis. To take on a fifth man as an afterthought could only be a grave risk. But Bowers had earned the supreme reward in that service of going to the Pole. *And Scott could not bear to disappoint him.*

Before Lieutenant Evans's party left, Scott wanted to see how the polar team could manage the full load. Bowers, on foot, was put between and behind Wilson and Scott, who were pulling on skis. At first it was doubtful whether he would be able to keep his own pace without throwing the other four out; but in a very short time it was obvious that all was well. Then, in Lat. 87° 34′ S. and 168 miles from the Pole, they halted and shook hands. 'Teddy Evans is terribly disappointed but has taken it very well and behaved like a man. Poor old Crean wept and even Lashly was affected.' The polar party bent forward in their harness, and Teddy Evans called for three cheers as they stepped off. They were last seen as a tiny black speck on the horizon.

Evans's party took back some letters, and amongst them was a line from Scott:

'A last note from a hopeful position. I think it's going to be all right. We have a fine party going forward and arrangements are all going well.'

* * * * *

On the flyleaf of Scott's diary. – Ages: Self 43, Wilson 39, Evans (P.O.) 37, Oates 32, Bowers 28. Average 36.

'A fine party going forward . . .'

Two months before, sixteen picked men had started south from Cape Evans, 750 miles away. Out of that company five had been chosen. Each of them had already shown uncommon stamina, so it is the more interesting to notice that three of them had at one time been invalids. Scott as a child had been the delicate subject of many anxious consultations; Oates had left Eton early on account of ill-health, and Wilson for the same reason had been obliged to live abroad for several years after taking his degree. All of them except Oates had ample sledging experience in low temperatures, and all except Petty Officer Evans were accustomed to taking responsibility and to leading others. None of them was likely to panic in emergency or to fail within the limits of human endurance.

Although earlier in the journey Scott had sometimes been very depressed, there can be little doubt that at this time he considered his position to be good. He had every confidence in his team, the weather was fine, and they appeared to be making excellent progress. His diary shows that he was now in far better spirits than he had ever been on the Barrier or the Beardmore. On the day that Lieutenant Evans, Crean and Lashly left them he wrote in his diary: 'With full marching days we ought to have no difficulty in keeping up our average. . . . At present everything seems to be going with extraordinary smoothness. . . .'

They travelled 49½ miles in the first four marches, but not without great difficulty, for the surface was very trying, being covered with ice crystals which hindered the gliding of the sledge. Scott considered these days to be the hardest that they had so far experienced on the plateau, but, in spite of the labour, he and his team were very far from being discouraged. The surface at this time was the only condition of which they complained. Shackleton's party had described the plateau to them as a truly awful place, and now they found that it was not nearly so bad as they had expected. '. . . the sun is so warm that in spite of the temperature we can stand outside in the greatest comfort. It is amusing to stand thus and remember the constant horrors of our situation as they were painted for us: the sun is melting the snow on the skis. . . . We feel the cold very little, the great comfort of our situation is the excellent drying effect of the sun. Our socks and finnesko are almost dry each morning.'

No, they were certainly not downhearted. Though they did not cease to revile the 'terribly heavy' surface they were uplifted by the knowledge that victory was in sight. '*What castles one builds now hopefully that the Pole is ours.*'

On one of these days – 5 January – there is an entry in Scott's diary which is easily overlooked but is of the greatest significance. 'Cooking for five takes a seriously longer time than cooking for four; perhaps half an hour on the whole day. It is an item I had not considered when reorganising.' Here is a great leader's recognition of a great blunder, which only one of the greatest of men could have so frankly admitted. For sheer magnanimity it is magnificent. Half an hour wasted a day: half an hour off the time allotted for sleep or half an hour off the day's march. Three and a half hours a week and two whole days by the middle of March. Just the difference between reaching One Ton Depot and failing to do so. It would be an overstatement to say that the addition of the fifth man wrecked the polar party: it was merely one of several unfavourable circumstances which together combined to defeat them.

On 7 January they camped at a height of 10,570 feet. This was the highest point reached, for after this they were travelling slightly downhill. They lay up for the next day and a half in a blizzard and, after the hard pulling of the four previous days, the rest was very welcome. They were in excellent spirits. 'Our food continues to amply satisfy. What luck to have hit on such an excellent ration! We really are an excellently found party.'

During this period of lying-up Scott wrote an appreciation of the other members of his team which must be quoted *in extenso*:

'It is quite impossible to speak too highly of my companions. Each fulfils his office to the party; Wilson, first as doctor, ever on the look-out to alleviate the small pains and troubles incidental to the work; now as cook, quick, careful and dexterous, ever thinking of some fresh expedient to help the camp life; tough as steel on the traces, never wavering from start to finish.

'Evans, a giant worker with a really remarkable headpiece. It is only now I realise how much has been due to him. Our ski shoes and crampons have been absolutely indispensable, and if the original ideas were not his, the details of manufacture and design and the good workmanship are his alone. He is responsible for every sledge, every sledge fitting, tents, sleeping-bags, harness, and when one cannot recall a single expression of dissatisfaction with any one of these items, it shows what an invaluable assistant he has been. Now, besides superintending the putting up of the tent, he thinks out and arranges the packing of the sledge; it is extraordinary how neatly and handily everything is stowed, and how much study has been given to preserving the suppleness and good running qualities of the machine. On the Barrier, before the ponies were killed, he was ever roaming round, correcting faults of stowage.

'Little Bowers remains a marvel – he is thoroughly enjoying himself. I leave all the provision arrangement in his hands, and at all times he knows exactly how we stand, or how each returning party should fare. It has been a complicated business to redistribute stores at various stages of reorganisation, but not one single mistake has been made. In addition to the stores, he keeps the most thorough and conscientious meteorological record, and to this he now adds the duty of observer and photographer. Nothing comes amiss to him, and no work is too hard. It is a difficulty to get him into the tent; he seems quite oblivious of the cold, and he lies coiled in his bag writing and working out sights long after the others are asleep.

'Of these three it is a matter for thought and congratulation that each is specially suited for his own work, but would not be capable of doing that of the others so well as it is done. Each is invaluable. Oates had his invaluable period with the ponies; now he is a foot slogger and goes hard the whole time, does his share of camp work, and stands the hardship as well as any of us. I would not like to be without him either. So our five people are perhaps as happily selected as it is possible to imagine.'

There was only one thing wrong – 'Evans has a nasty cut on his hand'. Scott hoped that it would not give trouble, but it must have already begun to do so or he would not have remarked on it; the accident had happened a week before, when the new runners were being put on, and until now had been considered too insignificant to merit a diary entry.

They got away after lunch on 9 January and covered 7 miles in a bad light but on a good surface. This took them to Lat. 88° 25′ S. – beyond Shackleton's farthest point. All was new ahead.

Eight days later, on 17 January, they reached the Pole; but they were fearful days of heavy toil, as the following analysis of Scott's diary shows:

10 Jan. 'Terribly hard march . . . surface is beyond words.'

11 Jan. 'never had such pulling . . . at fearful cost to ourselves . . . none of us have ever had such hard work before . . . a terribly trying time.'
12 Jan. 'worse than ever.'
13 Jan. 'very heavy dragging . . . tugging and straining.'
14 Jan. 'surface a little better.'
15 Jan. 'surface terrible.'
16 Jan.
17 Jan. The Pole. '. . . *an awful place*.'

That they found the pulling so heavy shows that the party was not so fit as it should have been. They were going downhill in the figurative as well as the literal sense: they were becoming worn out.

At the lunch camp on 10 January at One and a Half Degree Depot (*i.e.* $1\frac{1}{2}°$ of Latitude or 105 miles from the Pole) they built a cairn and left one week's food and various articles of clothing. This reduced the weight on the sledge by about 85 lb., and on 15 January they cached another four days' provisions 'and a sundry or two', – a further reduction of perhaps 50 lb. So the sledge at this time was comparatively light. The Pole was only a few marches ahead, and the triumph they would in a day or two achieve must have stimulated them. But, because they were weakening, they were hard-pressed to pull the lightened load.

On 12 January everybody was chilled while unlashing the sledge and putting up the tent. They imagined that the weather had turned colder, but, to their surprise, when they looked at the thermometer they found that the temperature was higher than on the previous night. That they suddenly felt the cold in that manner is another indication of their condition.

With the excitement of almost having reached their goal they did not realise that they were tiring. Indeed, on 14 January, Scott specifically stated that they were all very fit. When they felt the cold unexpectedly he attributed it to the exhaustion of the march and a damp quality in the air. When they found the hauling heavy they blamed the surface, and though it was far from good it cannot have been so bad as they thought.

They travelled 75 miles in the six marches from 10 January to 15 January, and then camped 30 miles from the Pole. That night Scott wrote: 'It is wonderful to think that two long marches would land us at the Pole. We left our depot today with nine days' provisions, so that it ought to be a certain thing now, and the only appalling possibility the sight of the Norwegian flag forestalling ours. Only 30[1] miles from the Pole. We *ought* to do it now.'

Next day the appalling possibility became an appalling fact.

They got away at 8 a.m. and had travelled 8 miles

1. Scott actually wrote '27 miles' because his unit of measurement was the geographical or nautical mile. For the purposes of this book it has been thought advisable to change all such reckonings to the more familiar statute mile of 1760 yards.

by 1.15 p.m. when they camped for lunch. At noon they had taken a meridian altitude which gave their Latitude as 89° 42′ S. So after lunch they started off in high spirits, feeling certain that the next day would see them at their destination. Bowers had the sharpest eyes, and was nearly always the first man to notice things. After they had gone about two miles he saw what he thought was a cairn away to the left, but managed to convince himself that it was one of the many irregularities in the wind-swept surface. Half an hour later he spotted a black speck ahead of them, and before they had gone much farther they had to face the fact that it could not be any kind of snow feature. They marched on and found a black flag; ski tracks and sledge tracks were all round it, and the foot-prints of many dogs. Sadly they pitched their tent.

'It is a terrible disappointment, and I am very sorry for my loyal companions. Many thoughts come and much discussion have we had. Tomorrow we must march on to the Pole and then hasten home with all the speed we can compass.'

FOUR

The Pole

(17–18 Jan. 1912)

'Great God! this is an awful place . . .'

They got under way at 7.30 a.m., and headed south in the teeth of a wind varying between 15 and 27 miles an hour, with the temperature *minus* 22° F. At 6 p.m. they camped in the region of the Pole. Wilson said it was the coldest march he had ever experienced. What a day to greet them!

'The Pole. Yes, but under very different circumstances from those expected . . . companions labouring on with cold feet and hands . . . the wind is blowing hard and there is that curious, damp, cold feeling in the air which chills one to the bone in no time . . . there is very little that is different from the awful monotony of past days. Great God! this is an awful place and terrible enough for us to have laboured to it without the reward of priority.'

And Wilson: 'It was difficult to keep one's hands from freezing in double woollen fur mitts. Oates, Evans and Bowers all have pretty severe frostbitten noses and cheeks, and we had to camp early for lunch on account of Evans's hands. It was a very bitter day.'

Because they were great men the bitterness was confined to the weather. The Norwegians had beaten them well and truly, so they bore no malice and made no uncharitable remarks. Instead, Scott paid them a tribute for the thorough way in which they had gone about their business and made sure of their work.

All through the previous winter, all the way across the Barrier, up the Beardmore and over the plateau to the Pole, they had considered the possibility of the Norwegians forestalling them. But when the blow came it staggered them. They had been sledging for 2½ months and were about 900 miles from home; they knew what it had cost them to reach the Pole, and how much more it would cost them to get back again. But it was worth it a thousand times over to discover what it is like at the end of the earth, where the sun never declines in the heavens, and where man always faces north whichever way he turns. But from the moment they saw that little black flag on 16 January they knew that they would have nothing fresh to tell the world. The whole journey, all those years of preparation and all those weeks of anxiety and courage and backbreaking toil were so much thistle-down and gossamer. No wonder they were shaken . . .

They slept that night, 17 January, at the Pole. To celebrate their arrival they had a double hoosh and some of the last bits of chocolate; they also enjoyed the cigarettes that Cherry-Garrard had so generously given them when they parted company. A final examination of all their observations made them think that they were $3\frac{1}{2}$ miles from the exact place – one mile beyond it and three to the right. So they struck camp at 5 a.m. and proceeded in that direction.

On the way they found a tent left by Amundsen. Above it the Norwegian flag was flying, and inside was the following record of five men having been there:

Roald Amundsen
Olav Olavson Bjaaland
Hilmer Hanssen
Sverre H. Hassel
Oscar Wisting. 16 Dec. 1911.

In addition to the record there was a letter addressed to King Haakon VII of Norway and a note for Captain Scott, as follows:

Poleheim,
15 December 1911.

'Dear Captain Scott,—As you probably are the first to reach this area after us, I will ask you kindly to forward this letter to King Haakon VII. If you can use any of the articles left in the tent please do not hesitate to do so. The sledge left outside may be of use to you. With kind regards I wish you a safe return.

Yours truly,
Roald Amundsen.'

There was also a quantity of clothing and equipment

in the tent, amongst which were three half reindeer-skin sleeping-bags, mittens, trousers, sleeping-socks, two sextants, two hypsometers and an artificial horizon. From this it appeared firstly that they had come prepared for far colder weather than they actually experienced, and secondly that they intended to follow a line of cairns for the return journey and did not anticipate doing much navigation.

Scott left a note saying that they had visited the tent; Bowers took some photographs, and Wilson made a sketch. Then they travelled on to the spot which from the previous night's observations they judged to be the actual Pole, and here they pitched their tent for lunch. Here also they built a cairn, hung Queen Alexandra's Union Jack and their sledging flags and took another photograph – 'mighty cold work all of it.'

Bowers took one final observation, and when it had been completed they walked a little farther on and there left a Union Jack flying. About half a mile away from this, the exact spot as near as they could fix it, they saw a black flag fixed to a sledge-runner. Wilson went over on skis and found a note signed by Amundsen attached to it. The splendidly accurate navigation of the two rival teams, who agreed in their location of the Pole, will always remain one of the greatest feats of the journey. When one thinks of observations taken in low temperatures with frostbitten fingers by worn out men at the end of long days of marching and sledge hauling, of chronometers subjected to all the jolts and hazards of pressure-ice, of spherical trigonometry computations worked out in a frozen sleeping-bag in a crowded tent, the feat becomes almost a miracle.

Scott gave Wilson the note and flag to keep. He had once said: 'I should like to have Bill to hold my hand when we get to the Pole.' While Bill had written: 'May I be there! About this time next year may I be there or thereabouts! With so many young bloods in the hey-day of youth and strength beyond my own I feel there will be a most difficult task in making choice towards the end.'

In the afternoon they travelled 6.2 miles towards the home they were never to reach.

'Well, we have turned our back now on the goal of our ambition and must face our 900 miles of solid dragging – and good-bye to most of the day-dreams!

'Now for the run home and a desperate struggle. I wonder if we can do it.'

FIVE

Return of the Supporting Parties

First Supporting Party	*Last Supporting Party*
Atkinson	Lieut. Evans
Cherry-Garrard	Lashly
Wright	Crean
Keohane	

THE FIRST Supporting Party, under the leadership of Surgeon-Lieutenant Atkinson, had turned back at the top of the Beardmore Glacier on 22 December. It is not possible within the compass of this work to give an account of their journey, and it must suffice to say that they reached Hut Point without undue difficulty on 26 January, having averaged as much as 16.3 miles a day until they reached One Ton Depot.

The return of the Last Supporting Party was an entirely different affair. Three-man units had done well before, but in this case two of the team had been in harness since the second motor broke down on 1 November. They left the polar party on 4 January, with the knowledge that they themselves would be hard pressed to cover the 750 miles back to Hut Point.

To reach their first depot on the return journey without going on short rations, they had to average 17 miles a day. In order to keep up this pace it was necessary to start travelling very early in the morning and to go on until late in the evening. To get an extra early start Evans would put the hands of his watch on an hour as soon as he woke up, and back again when he got an opportunity later in the day. This manœuvre was in order not to make Crean and Lashly unnecessarily anxious; but although the two seamen made no comment they were never for a moment deceived. All through the homeward journey they stole minutes from each day to add to their marches.

For three days they pushed on blindly in a blizzard, for they could not afford to lie up, and at the end of that time were some way off their course. However, eventually the sun reappeared and Evans was able to take an observation and correct the error. On 12 January they found themselves right above the perilously steep slope of the Shackleton Ice Falls. Coming south they had gone round them, but now food was running short so they considered going straight over; if they could descend 600–700 feet without smashing themselves to pieces they would save three vital days. The discussion was a matter of seconds, for the wind was cold and no time or place could have been less propitious for a quiet talk.

With sinking hearts they commenced the descent. The wind had swept away most of the snow, so the ice was bare and slippery. Every now and again the sledge took charge, fouled their traces and knocked them off their feet, but they dared not detach themselves for fear of falling down a crevasse. They were soon cut and badly bruised and had to clench their teeth to prevent themselves crying out with pain. At one point their speed must have approached 50 miles an hour; each man held on for his life, lying face downwards across the sledge. They shot over a yawning chasm before they knew of its existence, then crashed on an ice ridge beyond. The sledge rolled over and over, dragging them with it. How they arrived at the bottom without more serious injuries than scratches and contusions they could not understand.

They reached the Upper Glacier Depot on 14 January, with only one meal in hand, and here they collected the 3½ days' rations which were to take them down to the Middle Glacier Depot, 69 miles off; they stopped only as long as was necessary and then hurried away. On the 15th they covered nearly 22 miles, and about 19 on the following day. They camped in high spirits that night, thinking they were certain to reach their cache next day as it was only about 12 miles away; but they awoke to find that low clouds had filled the valley with mist. They set off down the glacier without waiting for the weather to clear, but before long got into bad pressure-ice, which became steadily worse; the harder they tried to get clear the more awkward did their situation become.

To return was unthinkable, so they struggled forward, at times almost carrying the sledge along. Where they could bridge crevasses with the sledge they did so, at other times they travelled along the ridges between until they came to a place where they could cross. Time and again they expected to be dashed down one of the fathomless pits they were skirting, and it was not until 11 p.m. on the day following that they won through to the depot.

'My God! what a day this has been for us all . . . no one would believe that we came through with safety, if we had only had a camera we could have obtained some photographs that would have surprised any one living. . . . Shall have reason to never forget the 17 and 18 January 1912. Tonight Mr Evans is complaining of his eyes, more trouble ahead!'[1]

They slept well on into the morning and then started off again. Lashly collected some geological specimens that afternoon – good work for a Chief Stoker. Evans's eyes were now very bad, and the two seamen took it in turn to lead. They reached the Lower Glacier Depot in three marches and felt quite proud of the performance. Next day, 22 January, they went down through the Gateway and saw the Barrier in front of them. That was the last of the Beardmore Glacier, and 'no more Beardmore for me after this,' wrote Lashly.

At Shambles Camp Evans realised that he had scurvy, for he felt stiff at the back of his knees, one of

1. Lashly.

the earliest symptoms. As Evans gradually got worse Lashly's responsibilities increased, and by 29 January he seems to have in some measure taken charge, for there is a note to the effect that he decided to stop Evans's pemmican that day. They continued to cover big distances every march. On 1 February Evans and Lashly had been out a hundred days, and Lashly celebrated it by changing his shirt inside out. He was wearing two shirts, and after this both sides of each would have done duty next to his skin. They always felt the benefit of such changes.

By 3 February Evans was unable to lift his legs, and the other two had to place him on his skis and strap him there. But he remained cheerful although obviously in great pain, and his pluck helped to encourage the two seamen.

Scurvy is a deficiency disease caused by the absence of a certain food property (now known as vitamin 'C') which is to be found almost exclusively in fresh meat, fruit, and vegetables. The melancholy narratives of many of the earliest expeditions show what fearful havoc this complaint caused amongst them. Captain Cook kept his crews free from it by giving them cress grown on board the ships on layers of cotton-wool, and it has been known for some time that fresh fruit juice is equally effective. However, this knowledge was not then of any help to men on sledging journeys, for they could neither carry the weight of fruit juice nor sit down and grow cress. All attempts to concentrate the juice had up to that time been unsuccessful, for this could only be done by heating it, and heat destroyed its magic.

On 4 February they reached the Upper Barrier Depot, which placed them 203 miles from Hut Point. At the rate at which they had so far been travelling this would only have taken them about fifteen more days. But it was obvious that they could not keep up this pace, for Evans was every day becoming a little weaker, and as he got worse so they got slower and slower. Lashly and Crean knew that if their return was much delayed Evans could not survive, and they hustled him along for all they were worth in the effort to get him back to fresh food in time. Before long they had to help him in and out of the tent and do everything for him. When he could no longer step out properly he contrived to waddle, pushing himself forward with a ski-stick and suffering agonies as he did so. Starting off one morning he fainted; Crean thought he was dead . . . and Evans came round to feel warm tears falling on his face.

On 13 February Evans collapsed completely and was quite unable to proceed farther. Crean and Lashly strapped him on the sledge in his sleeping-bag, with their own underneath his to make him more comfortable. He tried to persuade the others to leave him behind – 'but this we could not think of. We shall stand by him to the end one way or other. So we are the masters today. He has got to do as we wish and we hope to pull him through'.[1] Never was any man better served than Evans.

The weather was becoming very cold. Before starting out one morning Lashly changed his socks, and in doing so got one foot badly frostbitten, but Evans, weak as he was, brought it round again by placing it next to his stomach. 'I shall never forget the kindness bestowed on me in a critical time in our travels, but I think we could go to any length of trouble to assist one another; in such time and such a place we must trust in a higher power to pull us through.'[2]

Lashly and Crean struggled forward, pulling Evans on the sledge. But by this time they too were becoming exhausted, and their progress gradually got slower. They increased their hours of march, but this expedient was insufficient and on 16 February they went on half rations. On the following day they camped alongside the derelict motor, in which they were fortunate enough to find a few biscuits. They were now 34½ miles from Hut Point.

Next morning Evans fainted as soon as Lashly started to move him. 'Crean was very upset and almost crying, but I told him it was no good to create a scene but put up a bold front and try to assist.'[3] They brought him round with their last drop of brandy, and a little later managed to lift him on to the sledge and so began their march. But the surface had been spoilt by a recent

1. Lashly.
2. Lashly.
3. Lashly.

blizzard and they could make no progress. Their strength was spent.

Sadly they carried Evans back into the tent that was so nearly to be his grave, and then went outside again to discuss their next movements. They decided that Crean should try to reach Hut Point in order to obtain relief, while Lashly stayed to nurse the sick man. As there was only one day's provisions left, all Crean could take was three biscuits and two sticks of chocolate to sustain him on the way. The other two watched him go, wondering whether they would ever see him again.

Lashly was 'in a bit of a sweat' all that day, and remained up to watch the weather until long after midnight. It kept fine, which to some extent reassured him. Besides his doubts about the weather there was the fear that Crean might fall down a crevasse, and this risk was aggravated by the fact that he was not wearing skis, for these had been thrown away to save weight. After Crean had left, Lashly walked over to Corner Camp, only a mile away, and there found a note from Day to say that the crevasses on the way to Hut Point were particularly bad. To Evans Lashly merely remarked that he had found a note from Day, and that all was well.

A couple of days passed. The end had nearly come, but Evans was beyond caring. There was nothing left to eat except a few paraffin-saturated biscuits, and Lashly in his weakened state could never have marched in. He was forty-four years of age and due for his naval pension, a married man with a family. Cherry-Garrard has described him as an 'undefeated old sportsman', and certainly he took the present situation calmly enough. He had gone outside continually to look for their relief, but by the second evening it was almost blowing a blizzard, and they gave up all expectations for that night. Then they heard a dog whimper. They could scarcely believe their ears, but the sound was repeated, and they knew that help had come.

The dogs of the foremost sledge galloped right up to the tent, and the leader, a beautiful grey named Krisravitsa, seemed to understand the situation, for he came right into the tent and licked Evans's face and hands. As for Evans, he gave Krisravitsa the kisses that were meant for Lashly. Both of them were deeply affected by their rescue.

Crean started his walk to Hut Point at 10 a.m. on Sunday morning, 18 February, and did not stop until he had covered about 16 miles. He then sat down for five minutes and ate the chocolate and two of the three biscuits. He trudged on and reached the edge of the Barrier at about 12.30 a.m. on Monday, by which time he was getting very tired and cold. The weather was now a source of anxiety, for the sky had clouded over and a blizzard was imminent. He went down to the sea-ice which was slippery and gave him several falls on his back. It began to snow and the wind was rising, but he was too weary to go any faster. Somehow he did not lose direction, and staggered into Hut Point at 3.30 a.m.

Crean had covered the 34½ miles in eighteen hours. This was after a journey lasting three and a half months, for the last part of which he had been on short rations. Half an hour after he reached Hut Point the storm broke with full force. Had this happened any earlier nothing could have saved him, and the plight of his two companions would not have been known.

It was fortunate for Evans that the doctor and the dog-teams were at Hut Point when Crean arrived, and as soon as the blizzard abated they set out with all speed. Atkinson was very alarmed when he saw how far the scurvy had gone, and thought that Evans would not recover. However, he was able to give him immediately the fresh vegetables, fruit and meat for which his body was craving.

Dimitri had brought a large lump of cake for Lashly. 'It seems to me we are in a new world, a weight is off my mind, and I can once more see a bright spot in the sky for us all, the gloom is now removed.

'We are looking for a mail now. How funny we shall always be looking for something else, now we are safe.'[1]

1. Lashly.

SIX

Death of Petty Officer Evans

(19 Jan.—17 Feb.)

WHEN AT the Pole Scott wrote, 'now for the run home and a desperate struggle,' he must have realised the desperate plight they were in. It was at their 69th camp, and they had then been out 76 days from Hut Point. They could not hope to return very much faster than they had come. Their loads would be light, and for half the journey they would be travelling downhill; but this time they would have no ponies to help them across the Barrier, they were by no means so fit as when they had started, and the season was in decline. Seventy-six days from 18 January would take them on to 4 April. Shackleton was back at Hut Point on 28 February, but not before he had recorded a temperature of *minus* 34° F. So the polar party knew they would experience appallingly low Barrier temperatures, and they started for home wondering whether they would ever reach it. Seven days later Amundsen was on board his ship.

The 'pitiless, increasing wind' was now at their

backs, so they rigged up the floor-cloth of the hut as a sail. The mast was the staff of the black flag which had been their first intimation of the Norwegians' success. With a fresh breeze and full sail they travelled 18½ miles on 19 January, and next day picked up the depot which they had left four days previously. The red flag was flying merrily to welcome them back, and below it was the first of the three caches of food and fuel upon which they were dependent to get off the plateau alive. Until both the other cairns were picked up and the food safely on their sledge they could not be free from grave anxiety. Soon afterwards the wind strengthened and before long developed into a blizzard, which made them thankful that they had already retrieved their first depot.

If it had been hard work pulling downhill to the Pole, which was 800 feet lower than One and a Half Degree Depot, it could not now be any easier to pull uphill over the same surfaces. We find constant references to the exhausting nature of the work. The sheer foot-tons of energy required to haul sledge-loads to a height of 10,000 feet is prodigious, and the labour of continuing to haul them week after week without a break can only drain the body of its vital forces. They had reached the Pole by a superhuman effort. And now they were beginning to realise the price that had been paid.

Each man was feeling the cold more than on the outward journey, and Oates and Evans were becoming frostbitten. On 23 January Scott wrote: 'We came along at a great rate and should have got within an easy march of our depot had not Wilson suddenly discovered that Evans's nose was frostbitten – it was white and hard. We thought it best to camp at 6.45. Got the tent up with difficulty, and now pretty cosy after good hoosh. There is no doubt Evans is a good deal run down — his fingers are badly blistered and his nose is rather seriously congested with frequent frostbites. He is very much annoyed with himself, which is not a good sign. I think Wilson, Bowers and I are as fit as possible under the circumstances. Oates gets cold feet. One way and another, I shall be glad to get off the summit! We are only about 14 miles from our "Degree and Half" Depot and should get there to-morrow. The weather seems to be breaking up.'

They managed to cover seven of the fourteen miles next day before the wind became so strong that they had to camp. It was difficult work pitching the tent in a rising blizzard. Three men had to hold on to it for all they were worth while the other two piled snow on to the skirting to prevent it from blowing away. 'This is the second full gale since we left the Pole. I don't like the look of it. Is the weather breaking up? If so, God help us, with the tremendous summit journey and scant food. Wilson and Bowers are my standby. I don't like the easy way in which Oates and Evans get frostbitten.'

Next morning it was no use turning out at their

usual hour of 5.45 a.m., for the blizzard was as furious as ever. They decided to have breakfast later and to go without lunch. After a while the wind died down and they managed to get in a march and reach their cache. Three Degree Depot, 112 miles away, was now their only immediate source of anxiety.

The temperature was always between *minus* 20° F. and *minus* 30° F., and Wilson wrote: 'Our hands are never warm enough in camp to do any neat work now.' As well as having to endure cold and fatigue the party was getting daily more hungry. Though they hoped to increase the ration slightly if they arrived at the last summit depot in good time, they could not hope for a really full meal until they reached the pony-meat at the bottom of the Beardmore. 'A long way to go, and, by Jove, this is tremendous labour.'

From One and a Half Degree Depot it was downhill all the way to the bottom of the Beardmore Glacier, and they began to travel faster. Sometimes the wind drove them on with such speed that one or two men had to brake behind to keep the sledge from over-running the leaders. 'The sledge with our good wind behind runs splendidly . . . thank God the miles are coming fast at last.'

On 30 January they covered 22 miles for the second consecutive day, and camped only 5 miles short of Three Degree Depot. They were going down a slope over a good surface, the weather was fine and the wind favourable. But 'Wilson has strained a tendon in his leg; it has given pain all day and is swollen tonight. Of course, he is full of pluck over it, but I don't like the idea of such an accident here. To add to the trouble Evans has dislodged two finger-nails tonight; his hands are really bad and to my surprise he shows signs of losing heart over it.' They could get along with bad fingers, but if Wilson's lameness had got worse instead of better they would have been in an extremely awkward position. It is no exaggeration to say that any man who goes on a polar journey must face the possibility of having in certain circumstances to commit suicide in order to save his companions. A polar party must march or die. If a man develops appendicitis or breaks a leg in pressure-ice when far from home there is only one thing that he can honourably do.

Wilson walked by the side of the sledge next day and saved his leg as much as possible. They picked up the depot at midday and Bowers's skis in the afternoon. It will always be a mystery how Bowers had managed with his little short legs to slog along on foot for 450 miles, often in deep snow, doing his share of the hauling in the middle of four men pulling rhythmically on skis. No wonder Scott called him an 'undefeated little sportsman'!

They increased their ration by one-seventh and felt better for it immediately. 'The extra food is certainly helping us, but we are getting pretty hungry. . . . It is time we were off the summit. Pray God another four days will see us pretty well clear of it. Our bags are getting very wet and we ought to have more sleep.' Their sleeping-bags were their best friends. Inside them they could be quite alone and allow their thoughts to take them far away from sledging. Their hopes and all their little happinesses they could share with each other; but whenever one of them was bruised in spirit he could only turn his back on his companions, withdraw further into the privacy of his bag, and commune with himself. Before long the sleeping-bags had ceased to provide them with happy retreats, for ice formed wherever they breathed, and the bags had to be thawed out each night before they could pretend to be at all warm. When the bags were put outside in the morning they steamed in the cold air and had to be rolled up quickly before they froze stiff like a board. Not only did their breath condense and freeze in the sleeping-bags, but also upon the walls of the tent. By this time the inside of the tent was coated with a sheet of ice rising on all four sides from ground level to near the top. Small flakes of ice kept showering down, and if one of the party was unfortunate enough to rub up against the side in his sleep he would sometimes wake up in the morning to find himself frozen to the tent by his hair.

All the pleasure had gone from their sledging. They were feeling the pinch of hunger, they were in desperate need of rest, the tent and their sleeping-bags were lined with ice. And to add to their troubles Oates and Evans were far from well.

They were now approaching the crevasses above the Beardmore. Since leaving the Pole they had been retracing their steps along their outward route, but there was now a good landfall to help them find the Upper Glacier Depot, and on 3 February they decided to push on north without wasting time looking for tracks and cairns. They travelled 18 miles that day and 21 the next. Early on the second day they crossed several lines of crevasses deceptively covered over. Evans fell through twice and must have been a good deal shaken if not actually concussed. He began for the first time to show signs of failing mentally, and from that moment went steadily and rapidly downhill.

Evans was the biggest, heaviest and most muscular man in the party. He had to support an extra large body on food that was insufficient for his smaller companions, and, as Cherry-Garrard has pointed out, this must to a large extent have been the cause of his breaking down first.

There is no doubt that the condition of the party must have been giving anxiety, but it is impossible to say how much. They had great hopes of warmer temperatures on the Glacier and the Barrier, and for more food when they got to the pony-meat at Shambles Camp. But they were seriously worried about Evans, and in a lesser degree about Oates, who as early as 15 January had been feeling the cold and fatigue more than the others. Bowers – 'splendid, full of energy and bustle all the time' – and Scott were the fittest.

They reached the Upper Glacier Depot on 7 February, after three difficult days among pressure-ice. Mount Darwin, just west of the depot, was a good landmark to steer for, but although it was always in view it was not easy to reach. This was an anxious period, for all the time that they were threading their way through a maze of crevasses the weather was uncertain, and they could not afford to be held up as food was running short. However, they managed to reach the depot; and that, to their joy, was good-bye to the plateau. It had taken them 27 days to reach the Pole and 22 days to return – 7 weeks in low temperatures with an almost incessant wind. Just after they had left the Upper Glacier Depot on the outward journey Scott had written: 'We can pull our loads and pull them much faster and farther than I expected in my most hopeful moments.' But the team that had gone forward so strongly was now in a very different condition. The plateau had taken its toll.

They left the Upper Glacier Depot in a strong, cold wind, and made good speed downhill, Scott and Bowers in front on skis, with Oates and Wilson on foot on either side of the sledge. Wilson's leg was still not well enough for him to put on skis, while Evans was unable to pull and so was out of harness. At lunch-time the wind was still blowing half a gale, and they were all very cold and cheerless. But better things were in store for them. They decided to steer for a moraine under Buckley Island, and, pulling in crampons, they reached it after crossing some steep slopes with big crevasses. The moraine was obviously of great interest, and as they were now sheltered from the wind Scott decided to camp and spend the rest of the day studying its geology. They were overshadowed by perpendicular cliffs of Beacon sandstone through which ran several coal seams. From one of these Wilson picked up some plant impressions, including a piece of coal with beautifully traced leaves in layers. The weight of the geological specimens gathered on the way down the Beardmore Glacier was 35lb., a terrible additional burden for men in their condition with still so much farther to go.

There were 3½ days' rations at both the Upper and Middle Glacier Depot, which did not allow for any great delay among ice-falls. Going up the glacier they had been able to avoid the worst places, but coming down they could never see what was ahead, and time after time went over a slope to find themselves in the middle of very bad pressure-ice where it was impossible to tell which way to turn. Thus on the day after leaving the Upper Glacier Depot '. . . suddenly we found ourselves in pressure. Then came the fatal decision to steer east. We went on for six hours, hoping to do a good distance, which in fact I suppose we did, but for the last hour or two we pressed on into a regular trap. Getting on a good surface we did not reduce our lunch meal, and thought all going well, but half an hour after lunch we got into the worst ice-mess I have ever been in. For three hours we plunged on on skis, first thinking we were too much to the right, then too much to the left; meanwhile the disturbance got worse and my spirits received a very rude shock. There were times when it seemed almost impossible to find a way out of the awful turmoil in which we found ourselves. At length, arguing that there must be a way to our left, we plunged in that direction. It got worse, harder, more icy and crevassed. We could not manage our skis and pulled on foot, falling into crevasses every minute – most luckily no bad accident. At length we saw a smoother slope towards the land, pushed for it, but knew it was a woefully long way from us. The turmoil changed in character, irregular crevassed surface giving way to huge chasms, closely packed, and most difficult to cross. It was very heavy work, but we had grown desperate. We won through at 10 p.m., and I write after 12 hours on the march.'

They were uncertain of their exact whereabouts and still a fair way from the depot, which they did not expect to find without difficulty as it had been put down during a snowstorm. The pemmican hoosh was very small that night.

Next day, 12 February, they did a good march in the morning, and were relieved to see an old lunch camp, which showed that they were on the right track. They went on expecting to reach the depot without difficulty, but went too far to the left and got into another mass of pressure. By 9 p.m. they were in the worst place of all and too tired to go farther. 'After discussion we decided to camp, and here we are, after a very short supper and one meal only remaining in the food bag; the depot doubtful in locality. We *must* get there tomorrow. Meanwhile we are cheerful with an effort. It's a tight place, but luckily we've been well fed up to the present. Pray God we have fine weather tomorrow.'

They had fair weather next day, but not until they had been given cause for acute anxiety. Scott went outside several times in the night, only to find the sky clouding over and snow falling. When it was time to get up they could see nothing. They decided to have tea and a biscuit, but to keep the last meal of pemmican. They went forward by compass in the fog, threading their way through a turmoil of broken ice. Then the mist lifted and Wilson, to their immense relief, suddenly saw the depot flag. They camped at once for a meal.

Scott that night (13 February) wrote: 'Yesterday was the worst experience of the trip, and gave a horrid feeling of insecurity. Now we are right, but we must march. In future food must be worked so that we do not run so short if the weather fails us. We mustn't get into a hole like this again.' But two days later they were in similar difficulties. The depot was still about 20 miles away, and they had once again to reduce both food and sleep. None of them was pulling strongly, and they were becoming slow over camping arrangements. Evans was delaying them. For one reason and another they were taking almost as long to descend the glacier with a light load as it had taken them to climb it with a fully-laden sledge.

On 16 February they were obliged to camp early because Evans collapsed. No longer his normal reliable self, he stopped the march again and again on some trivial excuse until he was clearly unable to march farther. After a long night's rest he seemed better, and in the morning declared as usual that he was quite well. He took his place in the traces with the others, but had not put on his equipment properly, and his ski shoes soon worked loose. They waited for him to readjust them, but when it happened again and then again he was told to unhitch, get his footgear right, and then catch up the others. He did so in about an hour, and once more took his place in the team, but a little later he dropped behind on the same excuse. Scott cautioned him to come on as quickly as he could, and Evans answered cheerfully. They pulled on over a surface heavy with snow which had fallen the day before, and then camped for lunch, expecting Evans to be with them by the time that the water was boiling. But they finished their biscuit and tea, and still he did not come.

All four started back on skis. They found Evans on his knees, crawling. His hands were bare, his clothing was disarranged, and there was a wild look in his eyes.

The three fittest men returned for the sledge while

Oates sat with the dying man. What can his thoughts have been during that tortured vigil? He knew that his hands and feet were in much the same state as Evans's, and that he was by no means so well as the others. And he knew the impatience they all felt. At lunch that day they had for the first time openly discussed their desperate situation – over 400 miles from home, the season advancing rapidly, and a sick man upon their hands. Evans's death was their only chance.

They put him on the sledge and took him down to their tent. He died that night without recovering consciousness.

Edgar Evans, an extraordinarily good sledge-master and a sterling companion, collapsed from over-working and under-feeding. At home he would have gone to hospital; but a polar party must march or die. Evans pulled in harness on the day of his death, and when he could no longer march he endeavoured to crawl on his frostbitten hands and knees.

They buried him below the Monument Rock. Two hours later they packed the sledge in silence, and went down over the pressure ridges to the Lower Glacier Depot.

SEVEN

Death of Oates

(18 Feb.—16 Mar.)

THEY SLEPT for five hours at the Lower Glacier Depot, and then, at 2 p.m., went through the Gateway to Shambles Camp.

It was a badly shaken party that came to the last lap of 425 miles across the Barrier to Hut Point, but the ample supplies of pony-meat and the extra rations of Evans gave them renewed hope. If only their failing strength would allow them to keep up their distances they could win through.

The first few days were not reassuring. They did not leave Shambles Camp until noon on 19 February as there was a lot of work to be done, such as fitting the mast to a new sledge and digging out and packing up the pony-meat. They only travelled about 5 miles that afternoon, and not more than 24 miles during the first three marches, for they experienced the same trouble with the surface as on the plateau. In other respects things were improving. They had their full measure of food again, and managed to get their sleeping-bags a little drier by spreading them out on the sledge while they marched. Everyone agreed that horse-flesh cooked with pemmican made the best stew they had ever tasted.

Scott was hoping that travelling conditions would gradually improve as they got farther out on to the Barrier. In the lee of the mountains they were in a calm area. What they needed was a strong southerly wind, both to improve the surface and to fill their sails; and there was good reason to expect this wind, for Shackleton had it day after day at this stage of his journey. 'Everything depends on the weather. . . . There is little doubt we are in for a rotten critical time going home, and the lateness of the season may make it really serious.' So they watched the weather anxiously, hoping for wind and fearing a blizzard every time the sky tended to cloud over.

On 22 February they got a breeze from S.S.E., and immediately the snow began to drift and they lost the line of cairns. They were less than two days from the Lower Barrier Depot, and they could not be certain whether they were to the right or the left of the track, nor if the weather would clear in time. It was an uncomfortable moment, but 'it is satisfactory to note that such untoward events fail to damp the spirit of the party. Tonight we had a pony hoosh so excellent and filling that one feels really strong and vigorous again.' They had a worrying march next morning until just before lunch when Bowers's sharp eyes detected an old cairn. They camped at the end of the day only three miles short of the depot, which they reached without difficulty the following day. Wilson suffered a bad attack of snow-blindness as a result of looking for the tracks.

At the Lower Barrier Depot they received a cruel blow. All their stores were in order except what was perhaps the most important of them – the fuel. It had been exposed to extreme conditions of heat and cold, and there had been a leakage. Probably the oil had partially vaporised in the warmth of the sun and had escaped past the stoppers, the washers of which had perished in the extreme cold. But, whatever the reason, the polar party had now to go short. And it could not afford to do so.

It was 82½ miles to the next depot and, apart from fuel, they had ten days' rations. In point of fact they covered the distance in 6½ days, a magnificent performance in the face of their difficulties. But that week was the critical time in their fortunes. At the Lower Barrier Depot they still had a fighting chance. If when they reached the Middle Barrier Depot they were no less fit, if they found there all the fuel to which they were entitled, and, above all, if the weather would perhaps bring a little wind from the south and at all events get no colder, they would almost certainly win through. At the beginning of that fateful week Scott wrote: 'For the time anxieties are thrust aside . . . I don't know what to think. . . . It is a race between the season and hard conditions and our fitness and good food.' Three days later (27 February): 'It is a critical position. We may find ourselves in safety at next depot, but there is a horrid element of doubt.' And at the end of the week when they reached the Middle Barrier Depot only to find a further shortage of fuel in a temperature that had fallen below *minus* 40° F.: 'We are in a *very* queer street since there is no doubt we cannot do the extra marches and feel the cold horribly.'

On 27 February the temperature suddenly dropped more than 10° F., and Scott wrote that the situation had become critical. The minimum thermometer registered *minus* 37° F. during the previous night. Even now, had this been only a short spell of cold weather, the party might have got home. But the temperature never rose again. From that moment it was only a matter of days before the end, and that they managed to struggle on for another three weeks is almost incredible.

Misfortunes rarely come singly. At the Middle Barrier Depot they not only found a shortage of fuel, but Oates's feet were discovered to be badly frostbitten, and the weather became overcast so that they again lost the line of cairns. They began to suspect that they could not pull through, and it was not long before they knew this for certain.

Adversity seems to have dogged their steps from the first. It started with the loss of nearly half the ponies during the depot-laying journey of March 1911. Then came the long spell of bad weather below the Beardmore Glacier which caused a loss of time and an unwelcome consumption of food. When they reached the Barrier on their way home the following wind failed them as long as they were strong enough to profit by it; instead they found an unexpected severity of temperature and a shortage of fuel. The last blow was the blizzard which came down upon them when they were almost in sight of One Ton Depot. These were all risks incidental to the enterprise. That some of them would have to be met was to be expected; the fatal peculiarity of this case was that they all had to be met. Every chance turned against them, and the most splendid human effort was worn down by an accumulation of misfortunes.

They left the Middle Barrier Depot on 1 March. Next morning came the strong southerly wind for which they had been waiting so long, and they rigged their sail hoping for a great march. But even with this additional help they could not manage to cover 10 miles. The surface which had lately been very good was now coated with a layer of woolly crystals which prevented the sledge from gliding freely. Even with a strong wind and a full sail they could go no faster. 'God help us, we can't keep up this pulling, that is certain. Amongst ourselves we are unendingly cheerful, but what each man feels in his heart I can only guess.'

They were pulling, literally, for their lives, and all

the time with the knowledge that their marches were neither fast enough nor far enough. Yet each man managed to keep up a cheerful appearance. They were still a long way from the home they now realised they would never reach. They were overworked and underfed, with parts of them rotting as the gangrene from their frostbites spread. Dirty shapeless sledging clothes, over dirty distorted bodies that had been without a change of linen for months, hid a spirit that was unquenchable. What tremendous, enduring *guts* those men possessed!

Between 2 and 5 March they managed to cover just 10 miles each day, but after this there were no more double figures. On 5 March they camped 31 miles from the Upper Barrier Depot, with five days' food but only two days' fuel. 'Providence to our aid! We can expect little from man now except the possibility of extra food at the next depot. It will be real bad if we get there and find the same shortage of oil. Shall we get there? Such a short distance it would have appeared to us on the Summit! I don't know what I should do if Wilson and Bowers weren't so determinedly cheerful over things.'

Owing to their exhausted state they did not reach the Upper Barrier Depot until 9 March, and so had to make two days' fuel last for four. This meant eating their pemmican with only the chill off it; however, they pretended to prefer it like that, and made elaborate plans for a banquet when they reached home.

Oates by this time was nearly done for, and the lack of hot food further aggravated his condition. His feet were very bad, and he had been lame since 5 March. As the pain increased he grew more and more silent, but his spirits still occasionally rose in spurts. Of the others Wilson was feeling the cold most, mainly on account of his self-sacrificing devotion in doctoring Oates.

On 6 March Oates was unable to pull any longer. He walked painfully behind the sledge and sat on it when the others were searching for tracks. Scott wrote that he was wonderfully plucky and never complained, although suffering greatly. On 10 March, the day after they had reached the Upper Barrier Depot only to find a further shortage of fuel, Oates asked Wilson what chance he had and what he should do. He was already contemplating his great sacrifice. Kind-hearted Uncle Bill, to whom they all turned for advice, could only say: 'Slog on, just slog on.'

On 11 March Scott wrote: 'Titus Oates is very near the end. What we will do or what he will do, God only knows. We discussed the matter after breakfast; he is a brave fine fellow and understands the situation, but he practically asked for advice. Nothing could be said but to urge him to march as long as he could.' For several days Oates had been holding them back. On 6 March Scott had written: 'poor Soldier has become a terrible hindrance,' and on 10 March: 'poor Titus is the greatest handicap,' but nobody could tell him so. He had toiled so long and endured so much with them and they loved him dearly, so, although he asked for advice and they were signing their own death-warrant by so doing, they urged him to slog on.

Oates could not decide what he should do. Death would be easy. It was agony to live. But would he be dying in order to save his companions, or merely accepting weakly the easier way?

After the discussion on 11 March Scott ordered Wilson, who was in charge of the medicine case, to distribute the opium tablets, so that any one of them would have the means of ending his troubles if he so desired. When they camped that night they were 63 miles from One Ton Depot, with 7 days' food. But 7 miles a day was all they could hope to accomplish —'7 × 7 = 49, leaving us 14 miles short of our distance, even if things get no worse. Meanwhile the season rapidly advances.'

On 14 March the temperature at midday was as low as *minus* 43° F., that is seventy-five degrees of frost. The wind blew from the North, as cold as the death it was bringing. Wilson could not get his skis off for some time when they camped for lunch, so Scott and Bowers had to unlash the sledge and put up the tent almost unaided. It took them so long that they were all terribly cold when at last they got inside. They were by now very slow over all camping arrangements, and although the lacing of their fur-boots was of lamp-wick because it was the easiest material for frostbitten fingers, it took them literally hours to put on their footgear each morning.

At their lunch camp next day, 15 March, Oates said he could go no farther and asked to be left behind in his sleeping-bag. The others could not agree to this and persuaded him to struggle on for the afternoon march. That night he was worse, and they knew that the end was imminent. He went to sleep hoping not to wake.

'Should this be found I want these facts recorded. Oates's last thoughts were of his Mother, but immediately before he took pride in thinking that his regiment would be pleased with the bold way in which he met his death. We can testify to his bravery. He has borne intense suffering for weeks without complaint, and to the very last was able and willing to discuss outside subjects. He did not – would not – give up hope to the very end. He was a brave soul. This was the end. He slept through the night before last, hoping not to wake; but he woke in the morning – yesterday. It was blowing a blizzard. He said, "I am just going outside and may be some time." He went out into the blizzard, and we have not seen him since.

'We knew that poor Oates was walking to his death, but though we tried to dissuade him, we knew it was the act of a brave man and an English gentleman. We all hope to meet the end with a similar spirit, and assuredly the end is not far.'

EIGHT

The End

(17–29 March)

THE FOLLOWING passages are taken from Scott's diary:

'*17 March*. I can only write at lunch and then only occasionally. The cold is intense, *minus* 40° at midday. My companions are unendingly cheerful, but we are all on the verge of serious frostbites, and though we constantly talk of fetching through I don't think any one of us believes it in his heart.

'We are cold on the march now, and at all times except meals. Yesterday we had to lay up for a blizzard, and today we move dreadfully slowly. We are at No. 14 pony camp, only two pony marches from One Ton Depot. We leave here our theodolite, a camera, and Oates's sleeping-bag. Diaries, etc. and geological specimens carried at Wilson's special request, will be found with us or on our sledge.'

'*Sunday, 18 March*. Today, lunch, we are 24 miles from the depot. Ill-fortune presses, but better may come. We have had more wind and drift from ahead yesterday; had to stop marching; wind N. W., force 4[1], temp. *minus* 35°. No human being could face it, and we are worn out *nearly*.

'My right foot has gone, nearly all the toes – two days ago I was proud possessor of best feet. These are the steps of my downfall. Like an ass I mixed a small spoonful of curry powder with my melted pemmican – it gave me violent indigestion. I lay awake and in pain all night; woke and felt done on the march; foot went and I didn't know it. A very small measure of neglect and have a foot which is not pleasant to contemplate. Bowers takes first place in condition, but there is not much to choose after all. The others are still confident of getting through – or pretend to be – I don't know! We have the last *half* fill of oil in our primus and a very small quantity of spirit – this alone between us and thirst. The wind is fair for the moment, and that is perhaps a fact to help. The mileage would have seemed ridiculously small on our outward journey.'

'*Monday, 19 March*. Lunch. We camped with difficulty last night and were dreadfully cold till after our supper of cold pemmican and biscuit and a half a pannikin of cocoa cooked over the spirit. Then, contrary to expectation, we got warm and all slept well. Today we started in the usual dragging manner. Sledge dreadfully heavy. We are 17.8 miles from the depot and ought to get there in three days. What progress! We have two days' food but barely a day's fuel. All our

1. 15 miles per hour.

feet are getting bad – Wilson's best, my right foot worst, left all right. There is no chance to nurse one's feet till we can get hot food into us. Amputation is the least I can hope for now, but will the trouble spread? That is the serious question. The weather doesn't give us a chance – the wind from N. to N.W. and *minus* 40° temp. today.'

'*Wednesday, 21 March*. Got within 13 miles of depot Monday night; had to lay up all yesterday in severe blizzard. Today forlorn hope, Wilson and Bowers going to depot for fuel.'

'*Thursday, 22 and 23 March*. Blizzard bad as ever – Wilson unable to start – tomorrow last chance – no fuel and only one or two of food left – must be near the end. Have decided it shall be natural – we shall march for the depot with or without our effects and die in our tracks.'

'*Thursday, 29 March*. Since the 21st we have had a continuous gale from W.S.W. and S.W. We had fuel to make two cups of tea apiece and bare food for two days on the 20th. Every day we have been ready to start for our depot *13 miles* away, but outside the door of the tent it remains a scene of whirling drift. I do not think we can hope for any better things now. We shall stick it out to the end, but we are getting weaker, of course, and the end cannot be far.

'It seems a pity, but I do not think I can write more.

R. SCOTT.

'Last Entry: For God's sake look after our people.'

Oates had walked out to his death on 16 March, and they lay up all that day in a blizzard. Before starting off on the following morning they abandoned what they could to reduce their weights even further; but they still clung to their diaries and the geological specimens. On the Beardmore Glacier, with one man already broken down and the difficulties piling up ahead of them, they did not discontinue their scientific observations but stopped to examine the rocks and to hunt for fossils, nor did they ever know that the specimens they collected would one day solve the question of the age and past history of that part of the Antarctic Continent. Later, in their last terrible marches through intense cold and blizzard they still carried them, together with their notes and observations. They brought these back for humanity although they themselves must perish, instead of throwing everything away and thinking only of their lives.

In the three days between 17 March and 19 March they managed, somehow, to cover another $18\frac{1}{2}$ miles in spite of severe frostbites and a bitter wind which cut short their marches. When they pitched the tent for the last time they were almost *hors de combat*. They had reached a point 13 miles from One Ton Depot with two days' food and less than one day's fuel, but the gravity of their frostbites was far their most serious menace. Even in this extremity their great courage and determination might have forced them on to their depot had not the blizzard come down on the 20th.

The blizzard swept on . . . While waiting for the end, Scott wrote twelve letters to console the grief of those who were dear to him. And while the driven snow volleyed and drummed against the side of the tent, the dying leader wrote clearly and without erasure the message to the Public, with a calmly-weighed justification of the enterprise. All these, together with their bodies, were found by the search party eight months later.

'MESSAGE TO THE PUBLIC'

THE causes of the disaster are not due to faulty organisation, but to misfortune in all risks which had to be undertaken.

1. The loss of pony transport in March 1911 obliged me to start later than I had intended, and obliged the limits of stuff transported to be narrowed.

2. The weather throughout the outward journey, and especially the long gale in 83° S., stopped us.

3. The soft snow in lower reaches of glacier again reduced pace.

We fought these untoward events with a will and conquered, but it cut into our provision reserve.

Every detail of our food supplies, clothing and depots made on the interior ice-sheet and over that long stretch of 900 miles to the Pole and back, worked out to perfection. The advance party would have returned to the glacier in fine form and with surplus of food, but for the astonishing failure of the man whom we had least expected to fail. Edgar Evans was thought the strongest man of the party.

The Beardmore Glacier is not difficult in fine weather, but on our return we did not get a single completely fine day; this with a sick companion enormously increased our anxieties.

As I have said elsewhere, we got into frightfully rough ice, and Edgar Evans received a concussion of the brain – he died a natural death, but left us a shaken party with the season unduly advanced.

But all the facts above enumerated were as nothing to the surprise which awaited us on the Barrier. I maintain that our arrangements for returning were quite adequate, and that no one in the world would have expected the temperatures and surfaces which we encountered at this time of the year. On the summit in Lat. 85° 86° we had –20°, –30°. On the Barrier in Lat. 82°, 10,000 feet lower, we had –30° in the day, –47° at night pretty regularly, with continuous head wind during our day marches. It is clear that these circumstances came on very suddenly, and our wreck is certainly due to this sudden advent of severe weather, which does not seem to have any satisfactory cause. I do not think human beings ever came through such a month as we have come through, and we should have got through in spite of the weather but for the sickening of a second companion, Captain Oates, and a shortage of fuel in our depots for which I cannot account, and finally, but for the storm which had fallen on us within 13 miles of the depot at which we hoped to secure our final supplies. Surely misfortune could scarcely have exceeded this last blow. We arrived within 13 miles of our old One Ton Camp with fuel for one hot meal and food for two days. For four days we have been unable to leave the tent – the gale howling about us. We are weak, writing is difficult, but for my own sake I do not regret this journey, which has shown that Englishmen can endure hardships, help one another, and meet death with as great a fortitude as ever in the past. We took risks, we knew we took them; things have come out against us, and therefore we have no cause for complaint, but bow to the will of Providence, determined still to do our best to the last. But if we have been willing to give our lives to this enterprise, which is for the honour of our country, I appeal to our countrymen to see that those who depend on us are properly cared for.

Had we lived, I should have had a tale to tell of the hardihood, endurance, and courage of my companions which would have stirred the heart of every Englishman. These rough notes and our dead bodies must tell the tale, but surely, surely, a great rich country like ours will see that those who are dependent on us are properly provided for.

R. SCOTT.

Mrs Wilson, Lieutenant Bowers's sister, and Sir James Barrie have been kind enough to allow these three letters to be reproduced:

To Mrs E. A. Wilson

MY DEAR MRS WILSON, – If this letter reaches you Bill and I will have gone out together. We are very near it now and I should like you to know how splendid he was at the end – everlastingly cheerful and ready to sacrifice himself for others, never a word of blame to me for leading him into this mess. He is not suffering, luckily, at least only minor discomforts.

His eyes have a comfortable blue look of hope and his mind is peaceful with the satisfaction of his faith in regarding himself as part of the great scheme of the Almighty. I can do no more to comfort you than to tell you that he died as he lived, a brave, true man – the best of comrades and staunchest of friends.

My whole heart goes out to you in pity. – Yours,

R. SCOTT.

To Mrs Bowers

MY DEAR MRS BOWERS, – I am afraid this will reach you after one of the heaviest blows of your life.

I write when we are very near the end of our journey, and I am finishing it in company with two gallant, noble gentlemen. One of these is your son. He had come to be one of my closest and soundest friends, and I appreciate his wonderful upright nature, his ability and energy. As the troubles have thickened his dauntless spirit ever shone brighter and he has remained cheerful, hopeful, and indomitable to the end.

The ways of Providence are inscrutable, but there must be some reason why such a young, vigorous and promising life is taken.

My whole heart goes out in pity for you. – Yours,

R. SCOTT.

To the end he has talked of you and his sisters. One sees what a happy home he must have had and perhaps it is well to look back on nothing but happiness.

He remains unselfish, self-reliant and splendidly hopeful to the end, believing in God's mercy to you.

To Sir J. M. Barrie

MY DEAR BARRIE, – We are pegging out in a very comfortless spot. Hoping this letter may be found and sent to you, I write a word of farewell. . . . More practically I want you to help my widow and my boy – your godson. We are showing that Englishmen can still die with a bold spirit, fighting it out to the end. It will be known that we have accomplished our object in reaching the Pole, and that we have done everything possible, even to sacrificing ourselves in order to save sick companions.

I think this makes an example for Englishmen of the future, and that the country ought to help those who are left behind to mourn us. I leave my poor girl and your godson, Wilson leaves a widow, and Edgar Evans also a widow in humble circumstance. Do what you can to get their claims recognised. Goodbye. I am not at all afraid of the end, but sad to miss many a humble pleasure which I had planned for the future on our long marches. I may not have proved a great explorer, but we have done the greatest march ever made and come very near to great success. Goodbye, my dear friend. – Yours ever,

R. SCOTT.

We are in a desperate state, feet frozen, etc. No fuel and a long way from food, but it would do your heart good to be in our tent, to hear our songs and the cheery conversation as to what we will do when we get to Hut Point.

Later. – We are very near the end, but have not and will not lose our good cheer. We have had four days of storm in our tent and nowhere's food or fuel. We did intend to finish ourselves when things proved like this, but we have decided to die naturally in the track.

As a dying man, my dear friend, be good to my wife and child. Give the boy a chance in life if the State won't do it. He ought to have good stuff in him. . . . I never met a man in my life whom I admired and loved more than you, but I never could show you how much your friendship meant to me, for you had much to give and I nothing.

NINE

Epilogue

REGRETTABLY IT cannot be said that Scott was a great explorer. His first expedition had shown considerable promise, but he failed to learn its principal lesson, the limitations imposed by man-haulage, and that this had been underlined by Shackleton's experiences of 1908–9. He remained no less oblivious of the many tremendous contemporary journeys by dog-sledge in the arctic regions which were duly noted by Scott's rival, Amundsen.

Having decided to take dog teams to the Antarctic in 1910, Scott should have sent men to over-winter in North-West Greenland to choose them and learn to drive them. And having selected a cavalry officer to be in charge of ponies, he should have sent him to pick them. Instead he relied upon contractors to supply his dogs and ponies, with predictable results.

So, in 1912, Scott set out for the South Pole with four different kinds of transport: man-haulage (long out-moded), tractors (still in an experimental stage), unsuitable ponies, and only two dog teams. The first three failed him, which can hardly be considered surprising. Scott had made too many mistakes and, as we have seen, he paid the penalty.

Nevertheless Scott fully deserves to remain a national hero. He will always be remembered and not because he was a great explorer (which, indeed, he was not), nor because he reached the South Pole, nor yet again because he died a heroic death. His name lives on because he was a great and noble man. He did not abandon his two sick companions (Evans and Oates). There are more than a thousand pages written by Scott in the chronicles of the two expeditions that he led – his own book about the Discovery Expedition, and the diary he kept before and during the great sledging journey and which was published posthumously. Most of us will never meet as beautiful a character as that which is revealed unconsciously in these narratives. Reading them one cannot fail to see a clear picture of the man; it is not necessary to have served under Scott in order to know him. First of all he was essentially a leader. He had great energy and determination, and although he worked his men to breaking point when necessary he was always thinking of their welfare. This is not the conventional picture of an ideal leader. It is the sober statement of men who sledged with Scott. Over and over again they told how he was always considering other people. He was strong, immensely strong, both mentally and physically, but he was also sensitive and shy, kind and gentle. It is his humanity, his faith, his steadfastness, and above all his simplicity which make him stand so high among the great men of his age.

Scott's men reached the pole and died with their task accomplished. What is of far greater importance is the invincible courage, the unblenching fortitude that Scott and his companions displayed.

The geological specimens they brought back are beyond price, but even these are worth nothing in comparison with the manner in which they were secured. Fighting for their very lives with the remorseless forces of Nature, those men clung in every-increasing peril and weakness to the data it was their duty to secure, and thus snatched victory out of the jaws of death, to perish in the moment of accomplishment. That is what counts. Had the specimens proved to be worthless lumps of granite the glory of the achievement would have been in no way diminished.

The news of the disaster to the polar party staggered the whole civilised world. The British Empire mourned as if for a monarch, and indeed the passing of Queen Victoria was its only parallel in sorrow. Every nation expressed sympathy: kings and queens and senates paid their tributes. It was realised that the death of Scott was a loss not only to his country but to the whole world.

Glossary

Blizzard A high wind at low temperature accompanied by drifting, not necessarily falling, snow.

Crevasse A deep crack or rift in the ice of glaciers, shelf-ice, and other land-ice formations.

Finnesko The fur boots worn by the Laplanders. They are made of reindeer skin with the fur outside, and have not the hard sole of the ordinary boot.

Glacier A body of ice formed in a region of perpetual snow, which is moving slowly downhill or has ceased to do so.

Moraine Rock debris associated with a glacier.

Névé The compacted snow of a snow-field; a stage in the transition between soft loose snow and glacier ice.

Nunatak An insular hill or mountain surrounded by an ice-sheet.

Sastruga An irregularity formed by the wind on a snow plain. 'Snow-wave' is not completely descriptive, as the sastruga has often a fantastic shape unlike the ordinary conception of a wave.

Tide-crack A crack between the land-ice and the sea-ice which rises and falls with the tide.

TITLES IN THE NEW WINDMILL SERIES

Chinua Achebe: *Things Fall Apart*
Louisa M. Alcott: *Little Women*
Elizabeth Allen: *Deitz and Denny*
Margery Allingham: *The Tiger in the Smoke*
Michael Anthony: *The Year in San Fernando*
Enid Bagnold: *National Velvet*
Stan Barstow: *Joby*
H. Mortimer Batten: *The Singing Forest*
Nina Bawden: *On the Run; The Witch's Daughter; A Handful of Thieves; Carrie's War; Rebel on a Rock*
Rex Benedict: *Last Stand at Goodbye Gulch*
Phyllis Bentley: *The Adventures of Tom Leigh*
Paul Berna: *Flood Warning*
Pierre Boulle: *The Bridge on the River Kwai*
E. R. Braithwaite: *To Sir, With Love*
D. K. Broster: *The Flight of the Heron; The Gleam in the North*
F. Hodgson Burnett: *The Secret Garden*
Helen Bush: *Mary Anning's Treasures*
A. Calder-Marshall: *The Man from Devil's Island*
John Caldwell: *Desperate Voyage*
Albert Camus: *The Outsider*
Victor Canning: *The Runaways; Flight of the Grey Goose*
Erskine Childers: *The Riddle of the Sands*
John Christopher: *The Guardians; The Lotus Caves*
Richard Church: *The Cave; Over the Bridge; The White Doe*
Colette: *My Mother's House*
Alexander Cordell: *The Traitor Within*
Margaret Craven: *I Heard the Owl Call my Name*
Roald Dahl: *Danny, Champion of the World; The Wonderful Story of Henry Sugar*
Meindert deJong: *The Wheel on the School*
Peter Dickinson: *The Gift; Annerton Pit*
Eleanor Doorly: *The Radium Woman; The Microbe Man; The Insect Man*
Gerald Durrell: *Three Singles to Adventure; The Drunken Forest; Encounters with Animals*
Elizabeth Enright: *Thimble Summer; The Saturdays*
C. S. Forester: *The General*
Eve Garnett: *The Family from One End Street; Further Adventures of the Family from One End Street; Holiday at the Dew Drop Inn*
G. M. Glaskin: *A Waltz through the Hills*
Rumer Godden: *Black Narcissus*
Angus Graham: *The Golden Grindstone*
Graham Greene: *The Third Man* and *The Fallen Idol*
Grey Owl: *Sajo and her Beaver People*
John Griffin: *Skulker Wheat and Other Stories*
G. and W. Grossmith: *The Diary of a Nobody*
René Guillot: *Kpo the Leopard*
Esther Hautzig: *The Endless Steppe*
Jan De Hartog: *The Lost Sea*
Erik Haugaard: *The Little Fishes*
Bessie Head: *When Rain Clouds Gather*
Ernest Hemingway: *The Old Man and the Sea*
John Hersey: *A Single Pebble*
Georgette Heyer: *Regency Buck*

TITLES IN THE NEW

Alfred Hitchcock: *Sinister Spies*
C. Walter Hodges: *The Overland Launch*
Geoffrey Household: *Rogue Male; A Rough Shoot; Prisoner of the Indies; Escape into Daylight*
Fred Hoyle: *The Black Cloud*
Irene Hunt: *Across Five Aprils*
Henry James: *Washington Square*
Josephine Kamm: *Young Mother; Out of Step; Where Do We Go From Here?; The Starting Point*
Erich Kästner: *Emil and the Detectives; Lottie and Lisa*
Clive King: *Me and My Million*
John Knowles: *A Separate Peace*
D. H. Lawrence: *Sea and Sardinia; The Fox* and *The Virgin and the Gipsy; Selected Tales*
Marghanita Laski: *Little Boy Lost*
Harper Lee: *To Kill a Mockingbird*
Laurie Lee: *As I Walked Out One Mid-Summer Morning*
Ursula Le Guin: *A Wizard of Earthsea; The Tombs of Atuan; The Farthest Shore; A Very Long Way from Anywhere Else*
Doris Lessing: *The Grass is Singing*
C. Day Lewis: *The Otterbury Incident*
Lorna Lewis: *Leonardo the Inventor*
Martin Lindsay: *The Epic of Captain Scott*
David Line: *Run for Your Life; Mike and Me*
Kathleen Lines: *The House of the Nightmare; The Haunted and the Haunters*
Joan Lingard: *Across the Barricades; Into Exile; The Clearance*
Penelope Lively: *The Ghost of Thomas Kempe*
Jack London: *The Call of the Wild; White Fang*
Carson McCullers: *The Member of the Wedding*
Lee McGiffen: *On the Trail to Sacramento*
Wolf Mankowitz: *A Kid for Two Farthings*
Olivia Manning: *The Play Room*
Jan Mark: *Thunder and Lightnings*
James Vance Marshall: *A River Ran Out of Eden; Walkabout; My Boy John that Went to Sea*
David Martin: *The Cabby's Daughter*
J. P. Martin: *Uncle*
John Masefield: *The Bird of Dawning; The Midnight Folk; The Box of Delights*
W. Somerset Maugham: *The Kite and Other Stories*
Guy de Maupassant: *Prisoners of War and Other Stories*
Laurence Meynell: *Builder and Dreamer*
Yvonne Mitchell: *Cathy Away*
Honoré Morrow: *The Splendid Journey*
Bill Naughton: *The Goalkeeper's Revenge; A Dog Called Nelson; My Pal Spadger*
E. Nesbit: *The Railway Children; The Story of the Treasure Seekers*
E. Neville: *It's Like this, Cat*
Wilfrid Noyce: *South Col*
Robert C. O'Brien: *Mrs Frisby and the Rats of NIMH; Z for Zachariah*
Scott O'Dell: *Island of the Blue Dolphins*
George Orwell: *Animal Farm*
K. M. Peyton: *Flambards*
Philippa Pearce: *Tom's Midnight Garden*
John Prebble: *The Buffalo Soldiers*